STERLING
Education

AF758680

Essential Biology

Plants & Photosynthesis

3rd edition

Copyright © Sterling Education

Customer Satisfaction Guarantee

Your feedback is important because we strive to provide the highest quality educational materials. Email us comments or suggestions.

info@sterling–prep.com

We reply to emails – check your spam folder

All rights reserved. This publication's content, including the text and graphic images or part thereof, may not be reproduced, downloaded, disseminated, published, converted to electronic media, or distributed by any means without prior written consent from the publisher. Copyright infringement violates federal law and is subject to criminal and civil penalties.

3 2 1

ISBN-13: 979-8-8855727-9-8

Sterling Education materials are available at quantity discounts.

Contact info@sterling–prep.com

Sterling Education
6 Liberty Square #11
Boston, MA 02109

© 2026 Sterling Education

Published by Sterling Education

Printed in the U.S.A.

STERLING
Education

From the foundations of a living cell to the complex mechanisms of gene expression, *Essential Biology Self-Teaching Guides* are a comprehensive compendium of clearly explained texts to learn and master multifaceted biology topics.

These guides provide a detailed review of essential biological processes of living systems. Develop a better understanding of cell and molecular biology, mechanisms of cell metabolism, plants and photosynthesis, evolution and natural selection, ecology and population biology. Learn the principles of genetics, microbiology, classification and diversity, as well as structure and function of anatomical systems. Reinforce your learning by working through the practice questions and detailed explanations.

Created by highly qualified biology instructors, researchers, and education specialists, these books empower readers by helping them increase their understanding of biology.

We sincerely hope that these guides are valuable for your learning.

250911akp

Featured on

USA TODAY FOX CBS NBC

Essential Biology Self-Teaching Guides

Eukaryotic Cell & Cellular Metabolism

Molecular Biology & Genetics

Nervous & Endocrine Systems

Circulatory, Respiratory & Immune Systems

Digestive & Excretory Systems

Muscle, Skeletal & Integumentary Systems

Reproduction & Development

Microbiology

Plants & Photosynthesis

Evolution, Classification & Diversity

Ecology & Population Biology

Visit our Amazon store

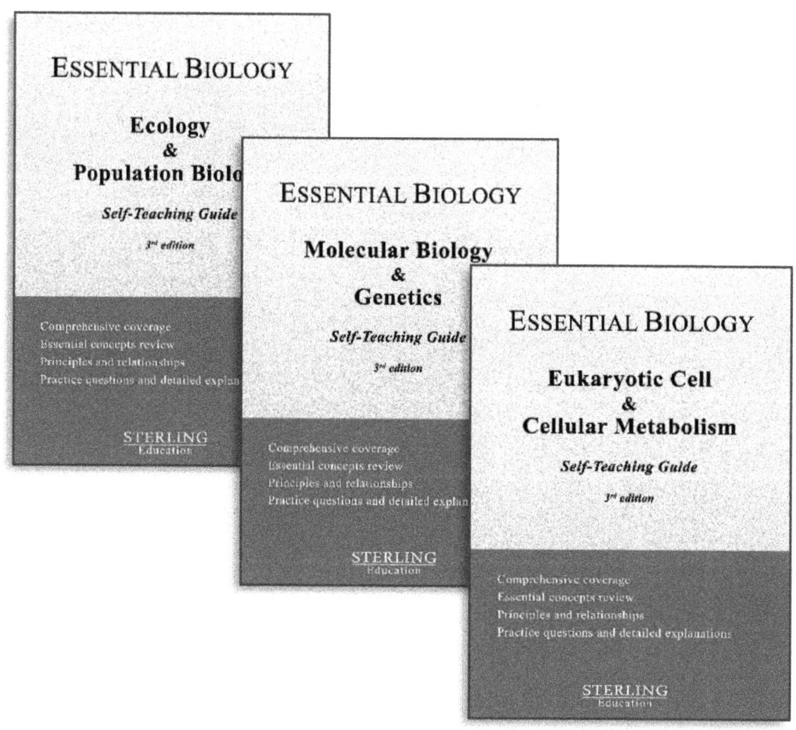

Copyright © Sterling Education

Essential Physics Self-Teaching Guides

Kinematics and Dynamics

Equilibrium and Momentum

Force, Motion and Gravitation

Work and Energy

Fluids and Solids

Waves and Periodic Motion

Light and Optics

Sound

Electrostatics and Electromagnetism

Electric Circuits

Thermal Physics

Atomic and Nuclear Physics

Visit our Amazon store

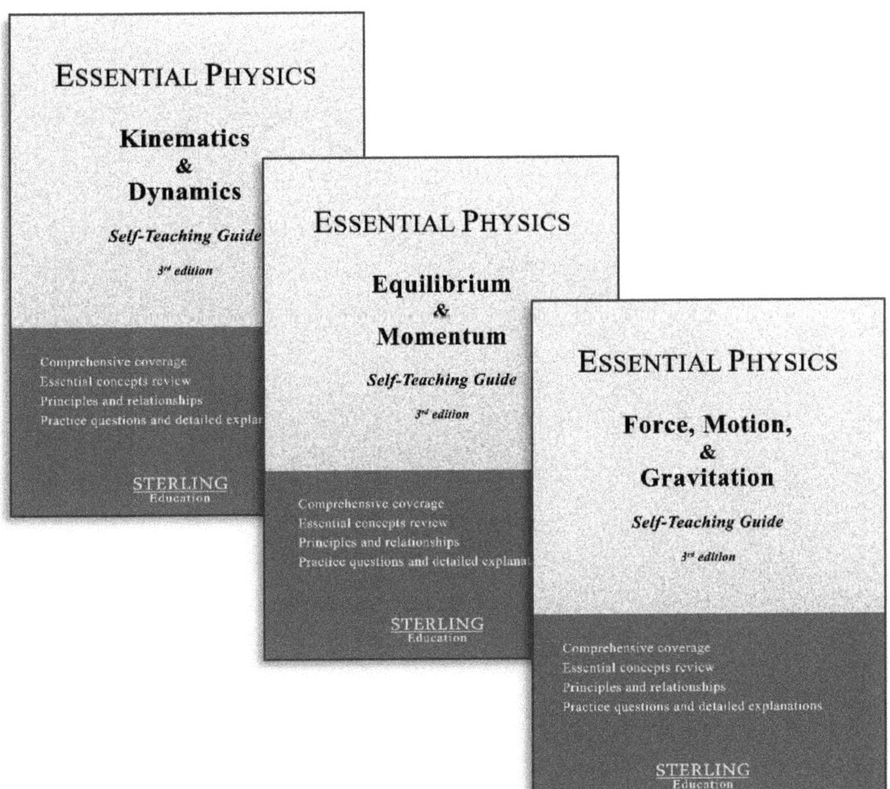

Essential Chemistry Self-Teaching Guides

Electronic Structure & Periodic Table

Chemical Bonding

States of Matter & Phase Equilibria

Stoichiometry

Solution Chemistry

Chemical Kinetics & Equilibrium

Acids & Bases

Chemical Thermodynamics

Electrochemistry

Visit our Amazon store

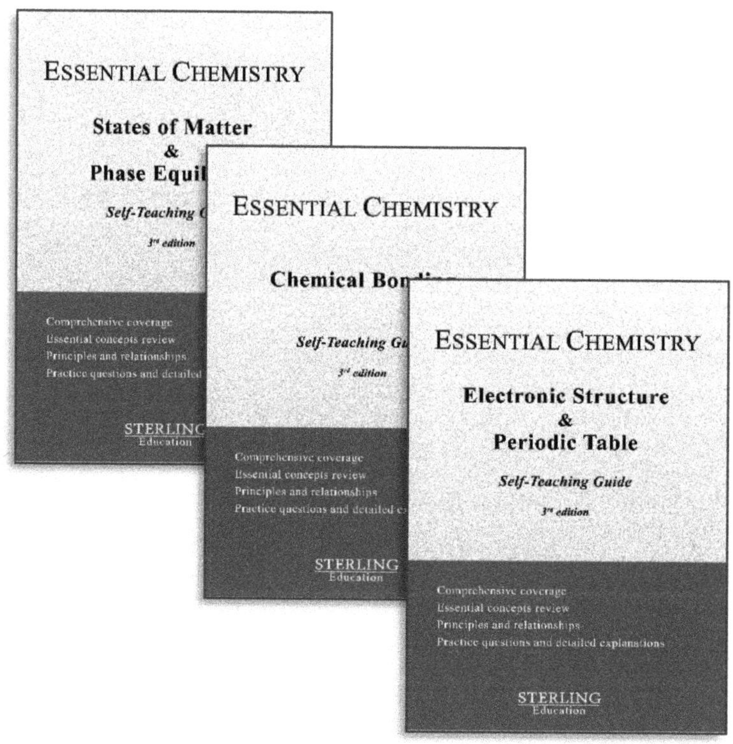

Copyright © Sterling Education

Everything You Always Wanted to Know About…

Chemistry

Physics

Cell & Molecular Biology

Organismal Biology

Human Anatomy & Physiology

Environmental Science

American History

American Law

American Government & Politics

Comparative Government & Politics

World History

European History

Psychology

Sociology

Human Geography

Visit our Amazon store

For online practice resources visit

https://www.sterling-prep.com

If you benefited from this book, please leave a review on Amazon so others can learn from your input. Reviews help us understand our customers' needs and experiences while keeping our commitment to quality.

Copyright © Sterling Education

Table of Contents

Any duplication (copies, uploads, PDFs) is illegal.

Table of Contents (*continued*)

REVIEW: Plants (*continued*)

 Copyright © Sterling Education

Table of Contents (*continued*)

Any duplication (copies, uploads, PDFs) is illegal.

Table of Contents (*continued*)

Copyright © Sterling Education

Table of Contents (*continued*)

Any duplication (copies, uploads, PDFs) is illegal.

Table of Contents (*continued*)

Copyright © Sterling Education

Table of Contents *(continued)*

Any duplication (copies, uploads, PDFs) is illegal.

Page intentionally left blank

Copyright © Sterling Education

REVIEW

Plants

Evolution and Diversity of Plants

Nonvascular Plants

Vascular Plants

Gymnosperms and Angiosperms

Vascular Plant Structure and Organization

Vascular Plant Growth

Specialized Plant Cells

Roots and Stems

Plant Nutrition

Transport in Plants

Control of Growth and Responses

Sporophyte and Gametophyte Generations

Reproduction in Plants

Seeds and Fruits

Page intentionally left blank

Copyright © Sterling Education

Evolution and Diversity of Plants

Evolutionary history of plants

Plants are multicellular, photosynthetic eukaryotes characterized under the kingdom *Plantae.*

The most primitive plants are *aquatic green algae*, the ancestors of all land plants.

Evidence for ancestry includes common adaptations by algae and terrestrial plants.

Adaptions include:

photosynthetic pigments,

storing *nutrients as starch* and

cell walls with cellulose, a tough polysaccharide.

Genetic analysis suggests that terrestrial plants' closest relatives are green algae, commonly known as *stoneworts*.

Green algae

Green algae (or *stoneworts*) colonized tidal areas and evolved into terrestrial plants in the *Paleozoic Era* as early as 600 million years ago.

Algae and early terrestrial plants are *nonvascular,* lacking a sophisticated vascular system.

Algae are more primitive and require a moist environment to maintain an adequate water supply for their tissues.

Nonvascular plants adapted to avoid desiccation with a *cuticle* (i.e., a waxy, protective layer).

However, the cuticle limits gas exchange, so plants developed *stomata* (openings) to allow CO_2 and oxygen to enter and exit their tissues.

Vascular tissue emerges

Around 430 million years ago, during the *Silurian Period.*

Nonvascular plants were well-adapted to drier environments and developed *vascular tissues* with specialized cells that conduct water and solutes throughout the plant.

Vascular tissue marks a significant division in the evolutionary history of plants, as it allows them to *venture inland* and *grow larger*.

Any duplication (copies, uploads, PDFs) is illegal.

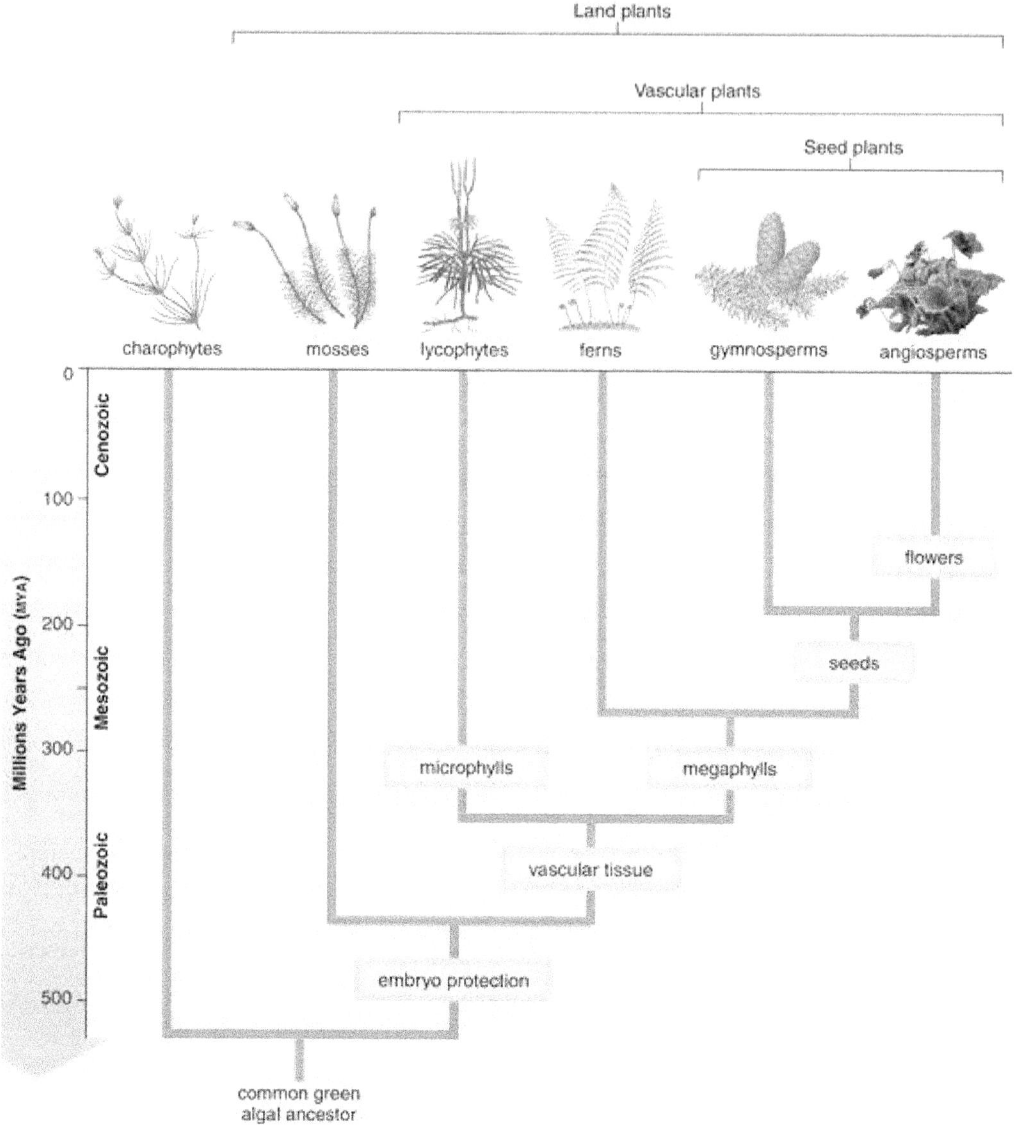

Evolution of plants from a common green algae ancestor

 Copyright © Sterling Education

Nonvascular Plants

Characteristics of nonvascular plants

Bryophytes	Pterophytes (and Others)
Nonvascular Plants	Vascular Plants (Seedless)
mosses, liverworts	ferns, horsetails, clubmosses
no vascular tissue	vascular tissue
no roots, leaves, stems	roots, leaves, stems
low lying	well developed fronds
spores	spores
gametophyte dominant	sporophyte dominant

Bryophytes and pterophytes with distinctive characteristics between species

Alternation of generations

Plants have a two-generation life cycle with *alternation of generations*, in which the morphology and reproductive strategy of the organism alternates.

Generations are named for the *offspring* they produce.

Sporophyte generation produces haploid *spores,* a unit of *asexual* reproduction.

Gametophyte generation produces haploid *gametes,* a unit of *sexual* reproduction.

Sporophyte translates to *spore plant*, while gametophyte translates to *gamete plant* (*phyte* is the Greek root for plant, *phyton).*

Any duplication (copies, uploads, PDFs) is illegal.

Gametophyte or sporophyte dominance

Gametophyte and sporophytes are *multicellular*, while their gametes and spores are *unicellular*.

Plant species differ in which generation, *gametophyte* or *sporophyte*, is dominant (i.e., more conspicuous), and the plant's appearance may vary widely between generations.

Gametophyte generation is generally dominant in *nonvascular plants.*

Sporophyte generation is generally dominant in *vascular plants.*

Adaptations for *water transport* and *conservation* accompany sporophyte dominance.

Sporophyte generation dominance in vascular plants is an adaptation to life on land and results in the gametophyte becoming microscopic and dependent on the sporophyte.

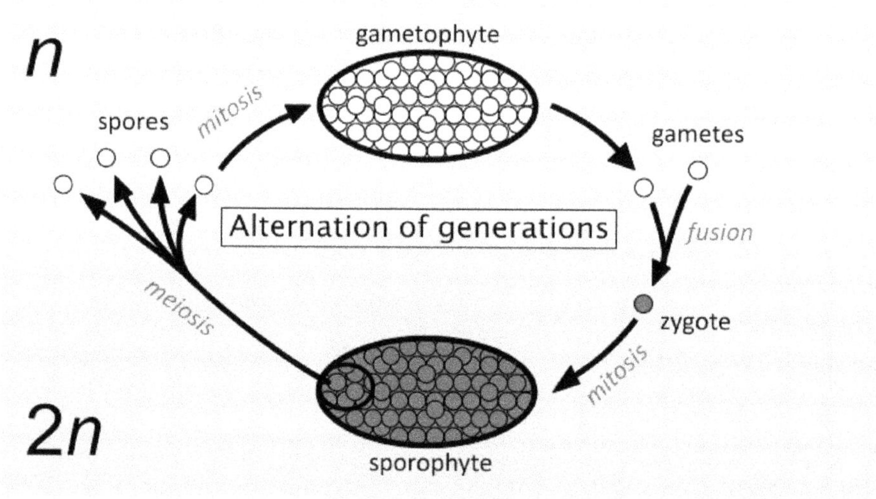

Alternation of generations with the 1N gametophyte compared to the 2N sporophyte

Nonvascular plants

Nonvascular plants lack true roots, stems, and leaves but may have substitute adaptations.

Nonvascular plants include *green algae* and *bryophytes*.

Nonvascular plants cannot grow large on land due to their lack of vascular tissue and usually require a moist habitat for fertilization.

Copyright © Sterling Education

Algae

Algae are aquatic, semi-aquatic, single-celled, colonial, or multicellular (e.g., seaweed).

Many algae species live symbiotically with fungi as *lichens.*

Unlike most plants, where one generation depends on the other, algae generally have sporophytes and gametophytes, both free-living.

Their sporophyte and gametophyte forms are practically indistinguishable.

Bryophytes are terrestrial nonvascular plants

Terrestrial nonvascular plants

Terrestrial nonvascular plants collectively known as *bryophytes* include three phyla:

hornworts (phylum *Anthocerotophyta*),

liverworts (phylum *Hepatophyta*) and

mosses (phylum *Bryophyta*).

Bryophytes

Bryophytes exhibit an alternation of generations in which the gametophyte dominates the sporophyte generation.

Like many of their algal cousins, bryophytes are small, low to the ground, and moist.

Bryophytes are anchored to the surface below by root-like *rhizoids.*

Like algae, many bryophyte species are capable of *asexual reproduction.*

The larger, dominant gametophyte form appears as a leafy or filamentous tissue mat from which small sprouts project.

Sprouts are male gamete-producing (*sperm*) or female organs (*eggs*).

Gametes combine within a female reproductive organ and develop into a sporophyte, which produces spores.

Spores are dispersed and develop into new gametophytes elsewhere, where the cycle continues.

Algae and bryophytes are *homosporous* (one type of spore).

Any duplication (copies, uploads, PDFs) is illegal.

Notes for active learning

Copyright © Sterling Education

Vascular Plants

Characteristics of vascular plants

Vascular plants are *homosporous* or *heterosporous* (i.e., two types of spores).

Vascular plants have a complex spore strategy, a hallmark of dominant sporophyte generations.

Gametophytes are tiny or microscopic, held within reproductive structures on the sporophyte.

Vascular plants have a *waxy cuticle*, *stomata* for gas exchange, and *vascular tissue.*

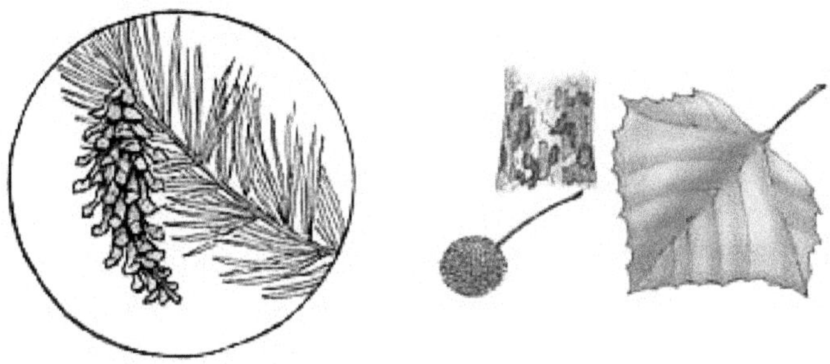

Gymnosperms	**Angiosperms**
Vascular Plants (Naked Seeds)	*Vascular Plants (Covered Seeds)*
conifers	*flowering plants, grass, and deciduous trees*
vascular tissue	*vascular tissue*
roots, leaves, stems	*roots, leaves, stems*
woody trees, needles	*trees, shrubs, broad leaves*
seeds	*seeds*
sporophyte dominant	*sporophyte dominant*

Gymnosperms and angiosperms divisions with distinctive characteristics

Seed and flower adaptions

Plants began to grow upright and quickly evolved *seeds* and *flowers* as adaptations for life on land around 400 million years ago.

Angiosperms are *flowering vascular plants* with reproductive structures to attract pollinators.

Gymnosperms are *non-flowering vascular plants* that use *cones* as reproductive structures.

Any duplication (copies, uploads, PDFs) is illegal.

Angiosperms and gymnosperms produce *seeds* containing food storage and a developing embryo within an outer coat.

Seeds have several advantages, such as drought and predator resistance, making the embryo more likely to mature successfully.

Seed evolution catapulted gymnosperms and angiosperms to the forefront of the plant kingdom.

Xylem and phloem vascular tissue

Xylem and phloem are two types of vascular tissues.

Xylem passively conducts water and mineral solutes, the *xylem sap,* from roots to leaves.

Water in the xylem rises against gravity without the help of any mechanical pump to reach heights of more than 100 meters in the tallest trees.

Phloem transports water, sucrose, and organic compounds, the *phloem sap,* through the plant.

Unlike xylem transport, phloem transport is active and thus requires energy to move its sap.

Seedless vascular plants

The first vascular plants were *seedless*, dominating the plant world from the late *Devonian* period (360 m.y.a.) through the *Carboniferous* period (300 m.y.a.).

Many plants were larger than their modern counterparts and formed vast, expansive swamps.

Like nonvascular plants, seedless vascular plants reproduce via *spores* rather than seeds and are often homosporous.

Seedless vascular plants include ferns, club mosses, whisk ferns, and horsetails.

Ferns

Ferns (class *Pteriopsida*) are the most well-known *seedless vascular plants.*

Ferns are the most diverse seedless land plant abundant in warm, moist, tropical regions.

Ferns may range in size from low-growing, moss-like forms to tall trees.

Ferns are the only seedless plants with well-developed *megaphylls*, which are "true leaves" with complex, branching veins.

Fern megaphylls are *fronds.*

Other seedless plants, as well as the lower plants (algae and bryophytes), have *microphylls* with only a single vein.

Many fern species live in the tropics or subtropics as *epiphytes,* plants that live on host trees without harming them.

Copyright © Sterling Education

Fern life cycle

Ferns are like vascular seed plants but have a life cycle with nonvascular characteristics.

Small gametophytes anchor to the soil with rhizoids and are free-living.

Fern gametophyte resembles a bryophyte, but most people would not recognize it as a fern.

In the rainy season, gametes produced on the underside of the gametophyte combine to create a diploid zygote, which matures into the adult sporophyte.

Sporophytes grow from the soil as large, leafy plants, typically known as *ferns*. It generates spores on the underside of its fronds.

Club mosses

Club mosses (lycopods or "ground pines") are not mosses but are another group of seedless vascular plants.

Club mosses are common in temperate woodlands, with tightly packed, scale-like microphylls covering their stems and branches.

Like ferns, club mosses have small, inconspicuous, free-living gametophytes and a more recognizable sporophyte.

Although seedless vascular plants may have independent gametophytes, the sporophyte is much larger and dominant.

Seed plants

Seed plants developed from seedless plants that were *heterosporous*.

Seed plants still exhibit heterospory, with small male and large female spores.

Male microspores are produced in the male reproductive organs.

Microspores develop into multicellular *microgametophytes* (male gametophytes), which produce sperm contained in *pollen grains.*

Meanwhile, the female megaspores are produced within *ovules* housed on the female plant.

Megaspores produced in ovules divide by mitosis into *megagametophytes* (female gametophytes).

Any duplication (copies, uploads, PDFs) is illegal.

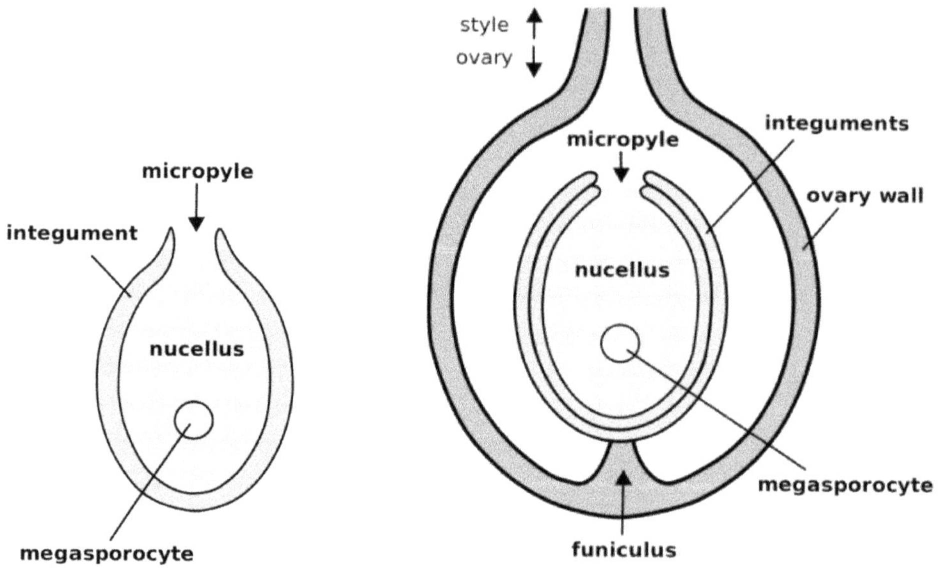

Gymnosperm (left) is a naked seed, and angiosperm (right) is an enclosed seed

When pollen fertilizes a macrogametophyte, it combines with the egg within to develop into a diploid zygote, which grows into an embryo contained in the seed.

Seeds

Seeds initially develop in ovules but are eventually ejected to germinate in soil.

Seeds are enclosed in a resilient protective coat for adverse conditions such as dryness and extreme temperatures.

Food is also stored in the seed to nourish the embryo until it can exist independently.

Seed plants' success and dominance are largely due to the utilization of seeds.

Copyright © Sterling Education

Gymnosperms and Angiosperms

Naked *vs.* enclosed seeds

Two divisions of seed plants are gymnosperms and angiosperms.

Gymnosperms have ovules *not* entirely enclosed by sporophyte tissue during pollination.

Angiosperms have ovules entirely enclosed within this tissue, the flower.

In ancient Greek, gymnosperm means "*naked seed,*" while angiosperm is "*enclosed seed.*"

Gymnosperms

Gymnosperms include conifers, cycads, ginkgo, and gnetophytes.

Gymnosperms have their reproductive organs partially exposed on *cones,* clusters of modified leaves with a scale-like appearance.

These spore-producing leaves are *sporophylls.*

Male cones are *pollen cones* because they produce the male gametophyte (pollen).

Female cones are *ovule cones* (seed cones) with ovules that house seeds upon fertilization.

Gymnosperms

Conifers

Conifers (Phylum *Coniferophyta*) contain about 575 species of cone-bearing trees and shrubs, such as pine trees, hemlocks, and spruces.

Usually, conifers have evergreen, needle-like leaves that are well-adapted to withstand extremes in climate due to their thick cuticles and recessed stomata.

Conifer forests cover vast areas of northern temperate regions and are characterized by being the oldest and largest trees.

Any duplication (copies, uploads, PDFs) is illegal.

Cycads

Cycads (Phylum *Cycadophyta*) have a stout and unbranched trunk with large, compound leaves that appear palm-like.

Cycads have pollen and seed cones on separate plants.

The cycad life cycle is like that of pine trees, except insects pollinate them.

Ginkgo

Only ginkgo survives from the phylum *Ginkgophyta* with fork-veined, fan-shaped leaves.

Ginkgo ovules reside at the end of short, paired stalks.

Gnetophytes (phylum *Gnetophyta*) comprise three living genera containing about 70 species.

Gnetophytes are the gymnosperms most closely related to angiosperms.

Some gnetophytes produce nectar in reproductive structures to recruit insects for pollination.

Angiosperms

Angiosperms (flowering plants) are the most numerous, with six times the species of all other plant groups combined.

Two hundred seventy-five thousand species of angiosperms live in habitats from freshwater to deserts and tropics to subarctic.

Angiosperms range from microscopic duckweed to *Eucalyptus,* exceeding 100 meters in height.

Unlike gymnosperms, angiosperms completely *enclose their ovules* within sporophyte tissues, specifically within the *ovaries* of flowers.

Fruits as mature ovaries

Fruits are *mature ovaries* that are often edible to attract organisms for dispersing seeds.

Seed dispersal is by wind, gravity, or water.

Vegetables are edible plant parts that are *not* part of the reproductive organs.

Copyright © Sterling Education

Angiosperm flowers

Angiosperm flowers have modified leaves and structures to facilitate *sexual reproduction.*

Receptacle is a modified stem tip that attaches the flower parts.

Sepals, a whorl of modified leaves protecting buds as the flower develops within, are outermost.

Petals are modified leaves interior to the sepals.

Petal shape and coloration account for the attractiveness of many flowers to pollinators.

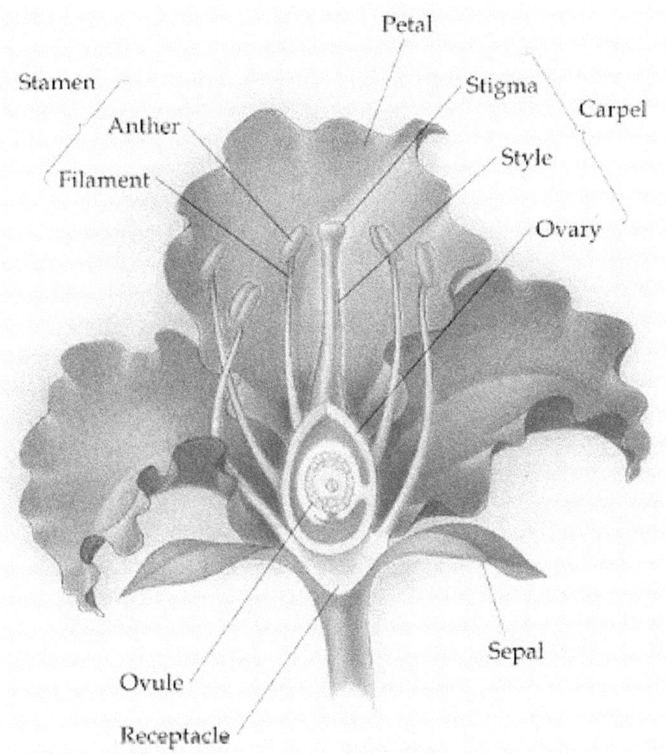

Flowers may have male or female reproductive organs or both

Angiosperm ovules are in a flower's ovary, while pollen grains are produced in *anthers.*

Glands in the ovary region produce *nectar* that nutrient pollinators gather among flowers.

The diversity of flowers is related to the numerous ways they are pollinated.

Color, size, shape, and scent all attract specific pollinators to disperse pollen.

Coevolution with insects and other animals encourages the dispersal of the pollen or fruit to locations suitable for germination and development.

Any duplication (copies, uploads, PDFs) is illegal.

Notes for active learning

Copyright © Sterling Education

Vascular Plant Structure and Organization

Root system

Vascular plant anatomy is divided into the *root* and *shoot* systems.

Root system consists of the primary root plus any side branches. It is generally equal in size to the aboveground shoot system and anchors the plant in the soil, providing support.

Roots absorb water and minerals from the soil, while *root hairs*, heavily concentrated near the root tip, further *increase the absorptive surface.*

Roots also produce hormones, which must then be distributed to the plant.

The roots of *perennials*, plants with leaves and stems that die at the end of each growing season, remain in the ground so that they may regrow next season.

Annuals die out at the end of the growing season and can only regrow from seed.

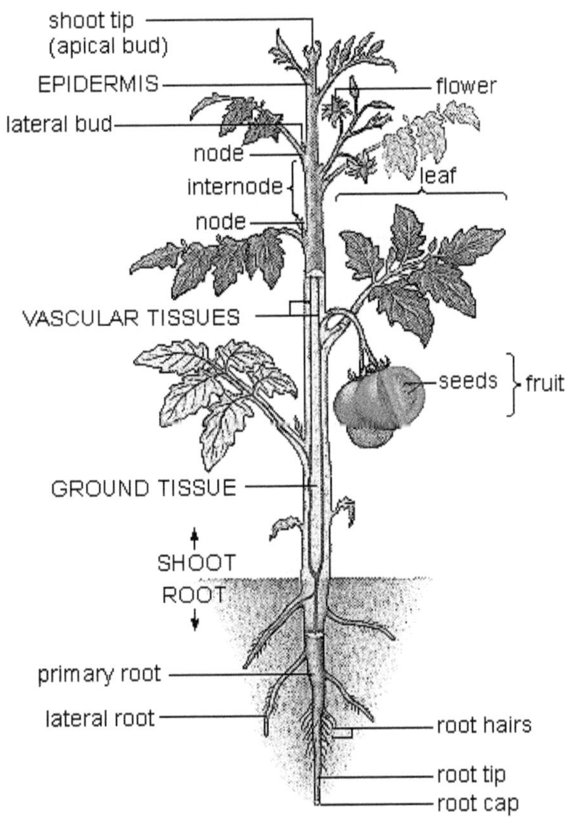

Angiosperm plant body: shoot system (above ground) and root system (underground)

Shoot system

Shoot system consists of the *stem*, *branches*, and *leaves.*

Stem forms the central axis of the plant, from which other plant organs branch out. Upright stems produce leaves and array them to be exposed to maximum sunlight.

Nodes are where a leaf attaches to the stem, and *internodes* are stem regions between nodes.

Some modified stems may be underground but can be distinguished by their nodes and internodes from the root system.

Like the roots, the stem has vascular tissue to transport water, organic nutrients, and inorganic minerals throughout the plant.

Trees use woody tissue to strengthen their stems and allow them to grow remarkably large.

Stems often function in the storage of water and nutrients.

Leaf is the major organ of photosynthesis in most plants. Leaves receive water from roots by the stem. Leaves are typically broad to maximize the surface area for CO_2 absorption and solar energy collection, and they are thin so that light can penetrate the leaf's underside.

Lamina (blade) is the wide leaf portion where most photosynthetic tissue is located.

Petiole is a stalk that attaches a leaf blade to the stem.

Leaf axil is the upper acute angle between the petiole and stems where *axillary buds* originate. Axillary buds may develop into additional leaves or flowers.

Leaves are modified for protection, structural support, and food storage.

Cotyledons

Two main groups of angiosperms are the *Monocotyledons* (monocots), with 65,000 species, and the *Eudicotyledons* (eudicots), formerly dicots, with 175,000 species.

Monocots produce one *cotyledon* at germination and often flower parts in multiple threes. Cotyledons are embryonic seed leaves nourishing embryos before and throughout germination.

Afterward, they fall off and are replaced by true, photosynthetic leaves or mature with the plant.

Cotyledon of monocots functions to transfer stored nutrients to the embryo.

Gymnosperm seedlings have cotyledons that are variable in number.

Eudicots produce two cotyledons at germination and often have flower parts in multiples of four or five. Cotyledons are nutrients themselves and directly feed the embryo.

Cotyledons of a gymnosperm are similar to monocots as they transfer nutrients to the embryo but do not feed it directly.

 Copyright © Sterling Education

Evolutionary division

The distinction between monocots and eudicots represents a crucial evolutionary division.

Monocots are palms, lilies, tulips, grasses, and grains (e.g., corn and wheat).

Eudicots are a larger group that includes oaks, maples, dandelions, cotton, and tomatoes.

In addition to the difference in floral arrangement and the number of cotyledons per seed, monocots and eudicots have other distinguishing characteristics.

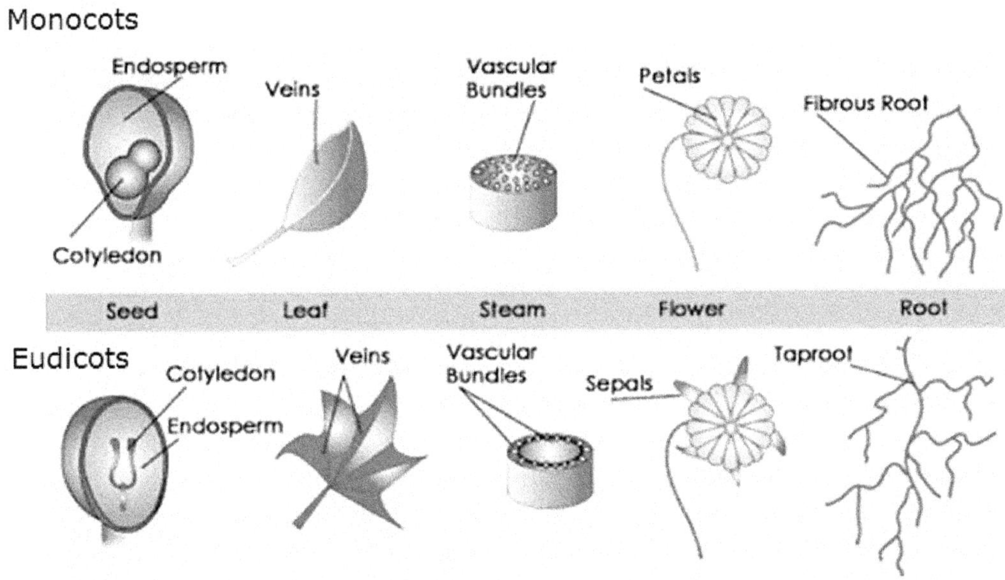

Monocot vs. eudicots distinctions

	Embryos	Leaf venation	Stems	Roots	Flowers
Monocots	One cotyledon	Veins usually parallel	Vascular bundles usually complexly arranged	Fibrous root system	Floral parts usually in multiples of three
Eudicots	Two cotyledons	Veins usually net-like	Vascular bundles usually arranged in a ring	Taproot usually present	Floral parts usually in multiples of four or five

Any duplication (copies, uploads, PDFs) is illegal.

Vascular tissue

Eudicots have vascular tissues arranged in a ring, while monocots it throughout the stem.

Nearly all monocots are *herbaceous* (relatively soft and pliable), while eudicots may be either herbaceous or *woody* (with rough bark).

Monocots typically exhibit parallel leaf veins, while eudicots have branched, net-like venation.

Vascular systems differ in the stems; monocots have *vascular bundles* of xylem and phloem randomly spread throughout the stem, while eudicots have their bundles arranged in rings.

Monocots exhibit a *fibrous root system* (or *adventitious root system*).

Fibrous root system consists of a mass of branching, slender roots originating from the base of the stem and holding the plant secure in the soil.

Conversely, eudicots begin life with a *taproot* (primary root system).

Primary root (or *taproot*) grows straight down and is the dominant root of the plant. It is large, fleshy, and central to any lateral branches.

Eudicots that retain a true taproot system throughout their lifetime include carrots, radishes, and beets. Most eudicots develop fibrous roots as they age.

Plant tissues

Plant and animal cells have several key differences. Unlike animal cells, plant cells have cell walls with a thick, rigid framework of *cellulose*, a tough polysaccharide.

Other polysaccharides, such as *lignin* for strength or *pectin* for elasticity, are in various concentrations depending on the cell type.

The walls shape the cell, regulate water and materials transport, and allow the plant to hold itself against gravity.

Plant cells connect by *plasmodesmata* (or channels of cytoplasm) to allow the passage of materials through the otherwise impenetrable cell walls.

Plant cells also contain *plastids* (organelles such as chloroplasts) not found in animal cells.

Vacuoles in plant cells are large and have a variety of functions, including carbohydrates, protein, and pigment storage. Vacuoles help maintain *turgor pressure*. Plant cells must remain *turgid* (swollen) to push the cell wall and maintain shape actively.

Without turgor pressure, cell membranes shrink from cell walls, causing them to be *flaccid* and compromising the plant's structural integrity. Wilting occurs from inadequate turgor pressure.

Vacuoles may or may not be present. As the cells mature, the vacuoles grow into different shapes and sizes, depending on the needs of the cell. Vacuole may fill 95% of the cell's volume.

 Copyright © Sterling Education

Vascular Plant Growth

Indeterminate growth

Plants experience *indeterminate growth*, as they do not have a pre-programmed body plan and continue to grow throughout their lives. There are constants like leaf shape and branching patterns, but it is impossible to predict precisely where a new branch develops on a tree.

Plants retain areas where rapidly dividing undifferentiated stem cells remain. Animals have stem cells but in smaller numbers and with limited *potency* (differentiation capability).

Plant stem cells are in *meristematic* tissue, which rapidly divides throughout the plant's lifetime to produce completely undifferentiated cells for growth and repair.

Meristem cells are generally small, six-sided, boxy structures with large nuclei.

Apical meristem and primary growth

Two regions of meristematic tissue are the *apical* and *lateral meristem.*

Apical meristem is located at or near the tips of roots and stems. This vertical growth is primary. As new cells form in the apical meristem, the plant elongates.

Root apical meristem lengthens the roots, while shoot apical meristem allows the formation of new leaves and flowers and lengthens the stem.

Vertical stem growth enables the plant to reach sunlight, while root lengthening is essential to anchor the plant and probe for water and nutrients in deeper soil layers.

While the apical meristem is in all vascular plants, a unique form, called *intercalary meristem,* is only in monocots such as grasses. These plants grow lengthwise from their tips via apical meristem and from their nodes via *intercalary meristem*. This allows grasses to regrow even when their tips are cut.

Primary meristem

Apical meristem gives rise to three *primary meristems,* each producing one *primary tissue*:

1) protoderm

2) ground meristems

3) procambium

Protoderm is the outermost primary meristem, which gives rise to *epidermal tissue.*

Ground meristem is the inner meristem that generates *ground tissue* as the inner bulk.

Procambium develops into *primary vascular tissue*, *primary xylem,* and *phloem.*

Any duplication (copies, uploads, PDFs) is illegal.

In some species, the procambium also gives rise to the *lateral* (or secondary) *meristems,* which account for secondary growth and are near the plant's periphery.

Lateral meristem and secondary growth

Secondary growth is generally horizontal in a direction (e.g., tree trunk in girth) and allows for extra xylem and phloem tissue production and stabilizes the plant to grow taller.

Primary meristems produce *primary tissues*, and lateral meristems produce *secondary tissues.*

Lateral meristems and secondary growth are in gymnosperms, woody eudicots, and a few woody monocots. These organisms still generally exhibit *apical dominance,* in which vertical or primary growth surpasses lateral or secondary growth due to hormonal control that promotes the action of *apical meristem* and inhibits *lateral meristem.*

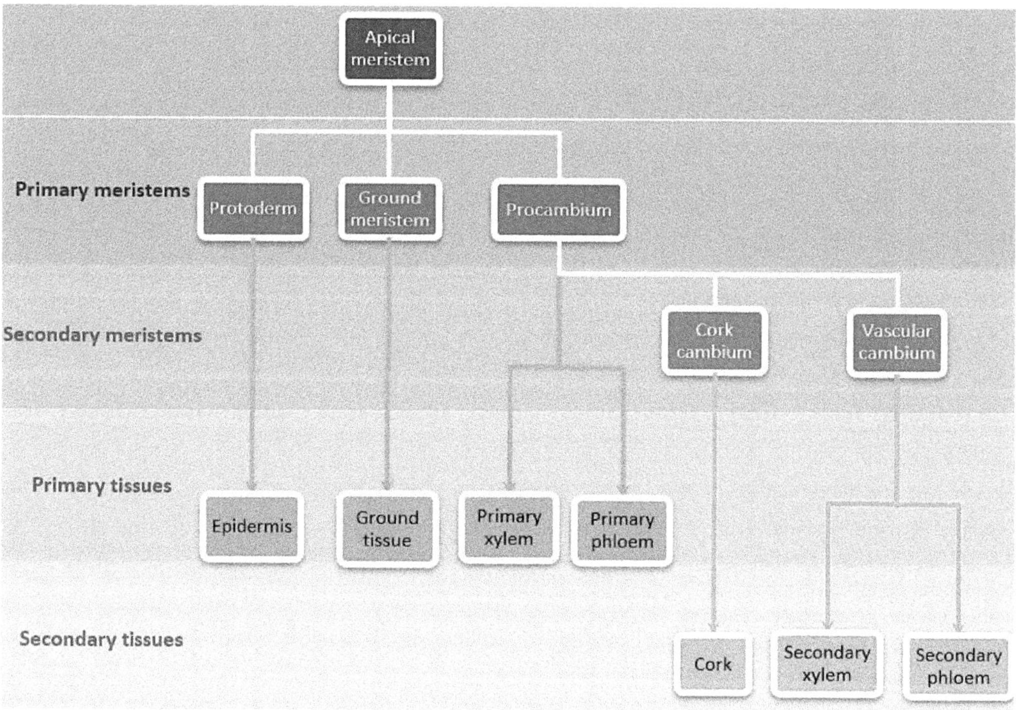

Apical meristem gives rise to three primary meristem and subsequent secondary meristems

Copyright © Sterling Education

Vascular and cork cambium

Vascular and *cork cambium*, two lateral meristems, are derived from *procambium*.

Vascular cambium is a thin, branching cylinder that runs the length of the roots and stems of most perennial and many herbaceous plants except for the tips with apical meristems. It produces the *secondary xylem* and *secondary phloem*, increasing girth.

Cork cambium is structurally similar to vascular cambium but is only in woody plants, where it produces the secondary tissue *cork*. As cork cells mature, they become encrusted with a waxy lipid called *suberin*, which renders them waterproof and inert, replacing the epidermis and producing outer bark.

Cork protects a plant and makes it resistant to attack by fungi, bacteria, and animals. If present, both the vascular cambium and the cork cambium begin to produce cells and tissues only after the primary tissues produced by the apical meristems have begun to mature.

Nonmeristematic tissues

Primary tissues (epidermal, ground, and primary vascular tissue) are *nonmeristematic.*

Nonmeristematic tissues are differentiated according to their intended function in the plant.

Tissues are composed of the same type of cells throughout or sometimes mixed.

Epidermal tissue is an outer protective covering for non-woody plants' roots, leaves, and stems. The epidermis is the outermost layer of tightly packed epidermal cells on all plant organs.

Generally, it is one cell layer thick, but there are exceptions, as in the case of tropical plants, where it may be several cells thick and act as a sponge.

In mature woody plants, the epidermis of the stem is replaced by bark. An epidermis is in direct contact with the environment and, therefore, is subject to environmental conditions.

Cuticle, created by a fatty substance called *cutin* secreted by most epidermal cells, covers the walls of epidermal cells to prevent water loss and protect against pathogens. Epidermal cells may be modified to serve various functions.

For example, specific epidermal cells in the roots are modified into root hairs to increase the surface area for water and nutrient absorption. Epidermal cells of stems and leaves produce hairs called *trichomes.*

Trichomes have a variety of functions, such as repelling insects and reducing water loss.

Any duplication (copies, uploads, PDFs) is illegal.

Ground tissue

Ground tissues comprise the bulk of the plant and consist of three specialized cell types:

1) parenchyma,

2) collenchyma, and

3) sclerenchyma.

Parenchyma has thin and flexible cell walls in all plant organs. Most parenchyma cells can differentiate into other cell types under special conditions, such as repairing and replacing organs after injury.

In leaves, some parenchyma cells differentiate into *chlorenchyma cells,* which have chloroplasts and form the *mesophyll tissue,* the plant's photosynthetic tissue.

Parenchyma cells have large vacuoles in storage organs that store starch, fat, protein, or water. They may also store resins, tannins, and crystals they secrete.

On the epidermis of leaves, each stoma is flanked by a pair of special parenchymal *guard cells.*

 Copyright © Sterling Education

Specialized Plant Cells

Guard cells and stomata

Guard cells are elongated and attached at ends, forming a donut shape. When they receive a signal to open stomata, they intake potassium ions, driving water to enter and increasing turgor pressure.

Guard cells enlarge in width because cellulose microfibrils restrict them but expand lengthwise, which causes the cells to buckle outward and create an opening between them.

Guard cells respond to changes in light, heat, and CO_2 concentration, as well as internal signals of water scarcity.

These factors are critical during plant development, as they affect the density of the stomata of the adult plant. CO_2 concentration rises in the air near the plant when photosynthesis ceases.

Evidence shows that this activates a receptor in the plasma membrane of guard cells, which shuts off the proton pump to close the stomata. Blue wavelengths of sunlight also trigger stomatal opening.

Pigments and hormones

Guard cells store *flavin pigments* in their vacuoles, which absorb blue light. Flavin pigments set a cytoplasmic response, activating proton pumps for potassium (K^+) ions to accumulate.

An internal biological clock may be responsible for this phenomenon because plants in the dark open and close their stomata like clockwork as if responding to the regular changes in sunlight over 24 hours. Plants actively control guard cells with hormones.

Abscisic acid (ABA) is a plant hormone with many uses, including initiating stomata closure.

Collenchyma and sclerenchyma cells

Collenchyma cells resemble parenchyma but have thicker primary cell walls, often with uneven thicknesses. They usually occur as bundles of cells just beneath the epidermis and are elongated.

Collenchyma cell walls are pliable and strong, making them well-suited to provide structural support in regions of the plant that are still growing.

The last category of ground tissue cells is the *sclerenchyma cells*, the strongest and least pliable.

Sclerenchyma cells may be extended and fibrous or more globular and irregularly shaped. They support mature plant regions and have thick secondary cell walls embedded with tough lignin.

Unlike other ground tissue cells, they are dead and cannot increase in length.

Any duplication (copies, uploads, PDFs) is illegal.

Primary xylem and phloem

Primary vascular tissue includes the primary xylem and phloem, which transport water, ions, minerals, and other nutrients and provide support.

Xylem and phloem extend from roots to leaves in bundles with supporting ground tissue.

Secondary xylem and phloem serve the same function as the primary ones. However, they are produced by lateral meristem during secondary growth rather than by the procambium during primary growth.

Typically, secondary vascular tissues are only in woody eudicots but may also be in a few monocots and herbaceous eudicots.

Xylem passively conducts water and mineral solutes (i.e., xylem sap) upward through a plant.

Cells of xylem tissue have two cell walls, primary and secondary.

Secondary cell walls are reinforced with lignin and are often deposited unevenly in various patterns to stretch. The cell walls are all left in xylem cells, as the cytoplasm and plasma membrane have disappeared. Therefore, the xylem is dead at functional maturity.

Vessel elements and tracheids

Vessel elements and *tracheids* are two types of xylem cells.

Vessel elements are hollow, non-living cells stacked end-to-end to reach from roots to leaves.

Vessel elements are connected vertically by *perforation plates* at their ends, allowing water and minerals to rise through a continuous pipeline.

Vessel elements are connected horizontally to other vessel elements by *pits* in their side walls.

Tracheids are hollow and non-living but are longer and slenderer than vessel elements. Columns of tracheids lie next to one another and reach throughout the length of the plant. They overlap at their tapered ends, with pits but *not* perforation plates.

Water rises through a tracheid until it reaches its end; at this point, it moves laterally through a pit into an adjacent tracheid and then rises until it reaches the next pit.

Tracheids form not a continuous pipeline but a "staircase" for sap transport.

Copyright © Sterling Education

Sieve-tube and companion cells

Phloem transports sucrose and other organic compounds from the areas where they are produced or stored (i.e., the leaves or storage organs) to any plant region requiring nutrients. It includes sieve-tube cells and companion cells.

Sieve-tube cells (sieve-tube members or sieve elements) are living cells arranged end to end and connected by plasmodesmata to form tubes. These porous connections between sieve tube cells (similar to the perforation plates between vessel elements) are *sieve plates.*

Sieve-tube cells live and have a cytoplasm actively involved in the conduction of food materials and do not have nuclei at maturity. Therefore, they depend on *companion cells* to sustain them.

Companion cells and vascular rays

Companion cells are specialized parenchyma cells intricately connected to sieve-tube cells by numerous plasmodesmata. They are smaller and more generalized than sieve-tube cells, with a nucleus that maintains the function of both cells.

Companion cells assist in transferring products from other plant cells to the sieve tube cells.

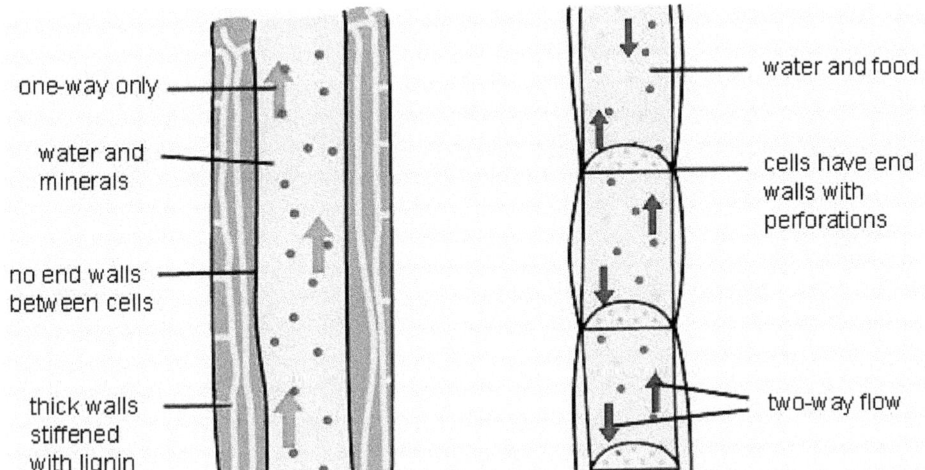

Xylem (left) and phloem (right) vessels

While most conduction in the vascular tissues is vertical, there is some side-to-side conduction via *vascular rays,* flat ribbons of parenchyma cells intersecting the vertical vascular system.

In a sense, vascular rays are the "*capillaries*" of the plant, transferring materials between the major arteries and veins of the plant (xylem and phloem) and the outer tissues.

In trees and woody plants, rays radiate from the center of stems and roots and look like the spokes of a wheel in cross-section.

Any duplication (copies, uploads, PDFs) is illegal.

Notes for active learning

Copyright © Sterling Education

Roots and Stems

Root and shoot tips

Root and shoot tips are sites of prolific primary growth due to the *apical meristem*. They are organized into zones of cells in various stages of differentiation.

Zone of cell division contains apical meristem and adds cells to the tip.

Zone of elongation is where cells become more specialized and undergo elongation, which pushes the root further down into the soil or shoots upward toward the light.

Zone of maturation (distal zone of elongation) is where cells are mature and differentiated.

Root cap protects the delicate apical meristem; its cells constantly slough off and are replaced.

Quiescent center is a population of cells in the apical meristem resistant to radiation and chemical damage, as a reserve if the apical meristem becomes damaged.

Flexible roots in non-woody plants (monocots, eudicots, and gymnosperms) are herbaceous.

Herbaceous eudicot root tissue is comprised of several cell types.

Vascular cylinder (or *stele*) is in the root center and contains xylem, phloem, and ground tissue.

Pericycle is the outermost layer of the vascular cylinder, through which all materials pass on their way to the xylem and phloem. Pericycle has lateral meristems and is responsible for secondary growth in *root branching*.

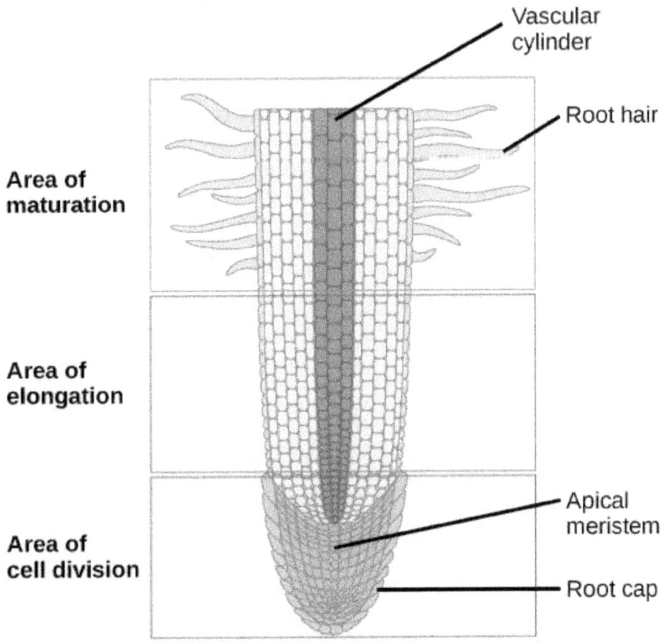

Root zone with root cap, apical meristem, and root hairs

Any duplication (copies, uploads, PDFs) is illegal.

Epidermis

Surrounding the vascular cylinder is a single ring of cells, the *endodermis.*

Its rectangular cells fit closely together and are sealed by the *Casparian strip*, an impermeable lignin and suberin layer that prevents water and minerals from entering the vascular cylinder by slipping between cells. Access to the vascular cylinder is by entering the endodermal cells.

Endodermis marks the boundary between the vascular cylinder and the *root cortex*, a layer of large, thin-walled, irregularly shaped parenchyma cells. These cells contain starch granules, as the cortex functions in food storage.

Cortex cells are loosely packed, allowing water and minerals to pass without entering the cells.

The outermost root ring is the epidermis, a single layer of thin-walled, rectangular cells that protect the root and extend root hairs for absorption.

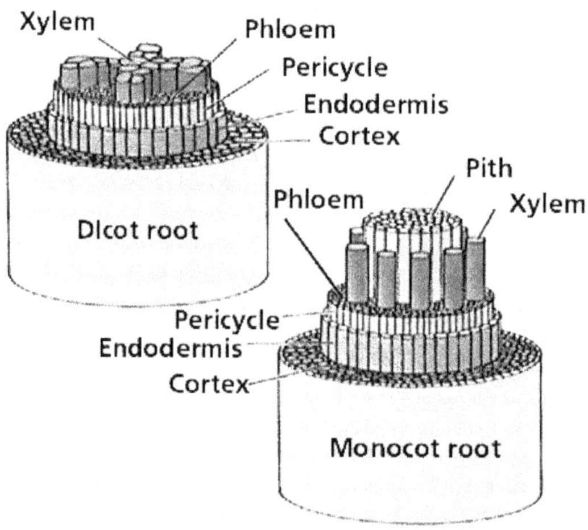

Herbaceous eudicot root and monocot root with an additional layer of pith vascular tissue

Monocot *vs.* woody roots

Monocot roots are nearly identical to herbaceous eudicot (dicot) roots but with an additional layer of central vascular cylinder, the *pith.* Xylem and phloem are arranged around this pith.

Woody roots have many more layers than herbaceous ones. While the outer layers (epidermis, cortex, and endodermis) are the same, the area within the endodermis is much more complex.

Endodermis surrounds a layer of cork produced by the cork cambium in the next interior layer. Cork makes these roots tough and woody. Moving from inside the cork cambium to the center, the rest of the root contains the primary phloem, secondary phloem, vascular cambium, secondary xylem, and primary xylem.

Vascular cambium continually divides to produce the secondary vascular tissues, which is why they are on either side.

Copyright © Sterling Education

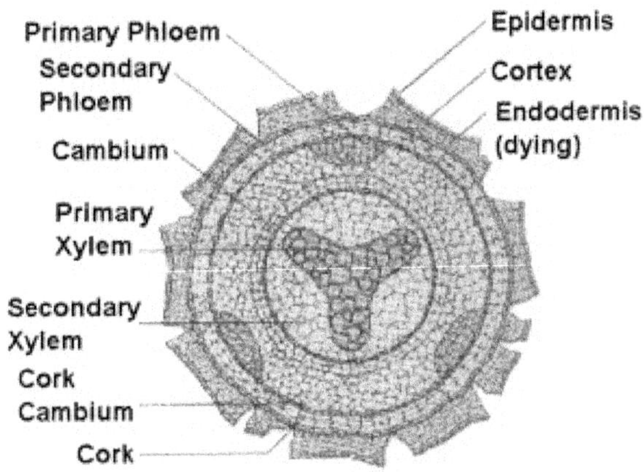

Woody eudicot / gymnosperm root

Woody monocot root is identical except for the pith within the primary xylem

Fibrous and tap root systems

Plants may have a *fibrous root system* or a *taproot system.*

However, the distinction is not entirely clear; a plant may have a taproot with lateral branches but also have fibrous roots originating from the stem or have a slender taproot with so many branches that it appears as a fibrous system.

Fibrous roots do not need to originate from a stem and can technically grow from any root.

Roots are heavily modified and do not exist in the underground root system.

Many roots are *aerial* (above ground).

Prop roots, for example, originate from the stem and reach the soil or other support structures to anchor a plant.

Pneumatophores project above the water to acquire oxygen. These roots are common in plants that live in swampy, aquatic environments, such as mangrove trees.

Haustoria is another form of aerial root, which projects from parasitic plants and latches onto the host plant to invade vascular tissues from which they extract water and nutrients.

Highly specialized *propagative roots* allow many plants to reproduce asexually. These roots bud off from horizontal stems called *stolons* that "run" on or under the soil, many feet away from the parent plant. The roots then develop into a clone of the parent.

Roots of many plant species develop mutualistic relationships with fungi and bacteria to maximize nutrient acquisition (discussed later).

Any duplication (copies, uploads, PDFs) is illegal.

Organization of stems

Some plants have adopted highly specialized stems that give them certain advantages.

Stolons stems grow horizontally along the ground, giving rise to new plants from their nodes.

Rhizome is a horizontal, underground stem like a stolon, but while stolons are small, thin stems that project from the main stem, a rhizome is the main stem itself.

Like stolons, some rhizomes have *stem tubers* that store nutrients, allowing the plant to survive winter, drought, and other adverse conditions.

Stem tubers serve as a means of asexual reproduction. An example of a stem tuber is a potato.

Other plants, such as cacti, have modified their stems for water storage.

Like prop roots, stems can be adapted to wrap around support structures. They are slim and provide an attachment and support, allowing the plant to climb upwards.

Like root tips, the stem tip is a site of primary growth where cell division extends its length.

Bud scales are scale-like coverings protecting dormant terminal buds in the winter.

Shoot apical meristem

Shoot apical meristem produces new leaves and primary meristems, increasing stem length. It is protected within a terminal leaf *primordia bud* (immature leaves).

Shoot apical meristem (like root apical meristem) gives rise to the three primary meristems:

1) protoderm

2) ground meristem

3) procambium

Monocot and herbaceous stems

Monocot stems have vascular bundles scattered throughout the stem; there is no well-defined cortex or pith.

Herbaceous stems have cross-sectional anatomy that features an epidermis with waxy cuticles surrounding ground tissues and vascular bundles.

Herbaceous stems of eudicots have vascular bundles arranged in a ring towards the outside of the stem, separating the outer cortex from the inner pith.

Herbaceous eudicot stems resemble monocot roots.

 Copyright © Sterling Education

Herbaceous eudicot stem with *central pith.*

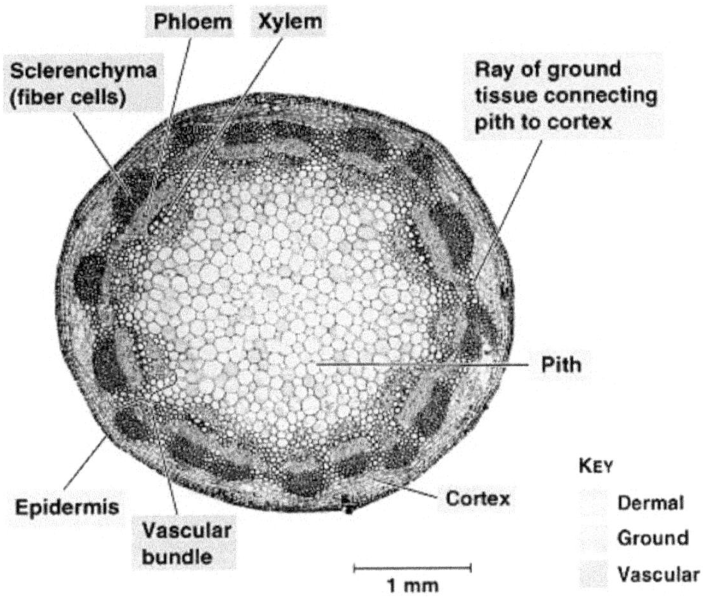

Herbaceous eudicot stems have vascular bundles arranged in a ring towards the outside, separating the outer cortex from the inner pith

Monocot stem with *ground tissue.*

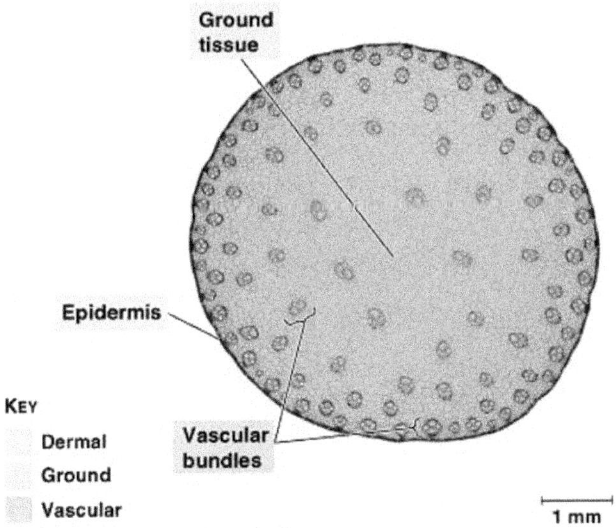

Monocot vascular bundles are scattered in stems with ground tissue and no well-defined cortex or pith

Any duplication (copies, uploads, PDFs) is illegal.

Cell growth

Like a woody root, a woody eudicot stem is more complex than herbaceous. Woody plants undergo only primary growth in their early years and are herbaceous. However, it eventually enters a secondary growth stage when the vascular cambium layer between the *primary phloem* and *primary xylem* produces new cells.

New cell growth forms a *sandwich* of primary phloem, secondary phloem, vascular cambium, secondary xylem, and secondary phloem.

Secondary phloem is soft and does *not* build up, while *secondary xylem* accumulates as *wood.*

Vascular cambium is dormant during the winter, so no wood is produced at that time.

Secondary xylem accumulates as wood

Springwood comprises broad xylem vessel elements with thin walls necessary to conduct sufficient water and nutrients to sustain abundant growth during spring.

Summerwood forms when moisture is scarce; composed of a lower proportion of vessels, it contains thickly-walled tracheids and numerous fibers.

Annual ring is a ring of spring wood followed by a ring of summer wood (i.e., 1 year's growth).

Sapwood is the outer annual ring of wood where vascular transport occurs.

Heartwood is made of the inner annual rings, found only in older trees.

Vessels in heartwood no longer function in transport; they have become plugged with resins and gums that inhibit the growth of bacteria and fungi.

Pith inside the vascular cylinder is often replaced entirely by *heartwood.*

Cork cambium, just interior to the epidermis, produces cork cells impregnated with *suberin,* which grow outward and eventually replace the epidermis. They die, forming an impervious barrier to gas and water exchange, physical damage, and pathogens.

Lenticels raised areas of loosely packed cells are the only areas allowing gas exchange.

Cork cambium

Cork cambium also sometimes produces a layer of living parenchyma cells.

Cork cells, cork cambium cells, and parenchyma cells, in combination, are the *periderms.*

After weeks, the cork cambium loses meristematic ability, and expansion splits the original periderm. New cork cambium then forms deeper in the cortex of the stem.

 Copyright © Sterling Education

Bark

All tissues exterior to the vascular cambium are considered *bark*. Bark includes the periderm, primary phloem, and secondary phloem from outside to inside.

It is advantageous to be woody when there is adequate rainfall, as woody plants can grow taller and larger if they can sustain their tissues.

Woody plants are longer-lasting and need more defense mechanisms against herbivores and parasites. They also mature much more slowly and thus cannot reproduce as quickly.

Before they mature, they may be destroyed by attack or disease before they reproduce.

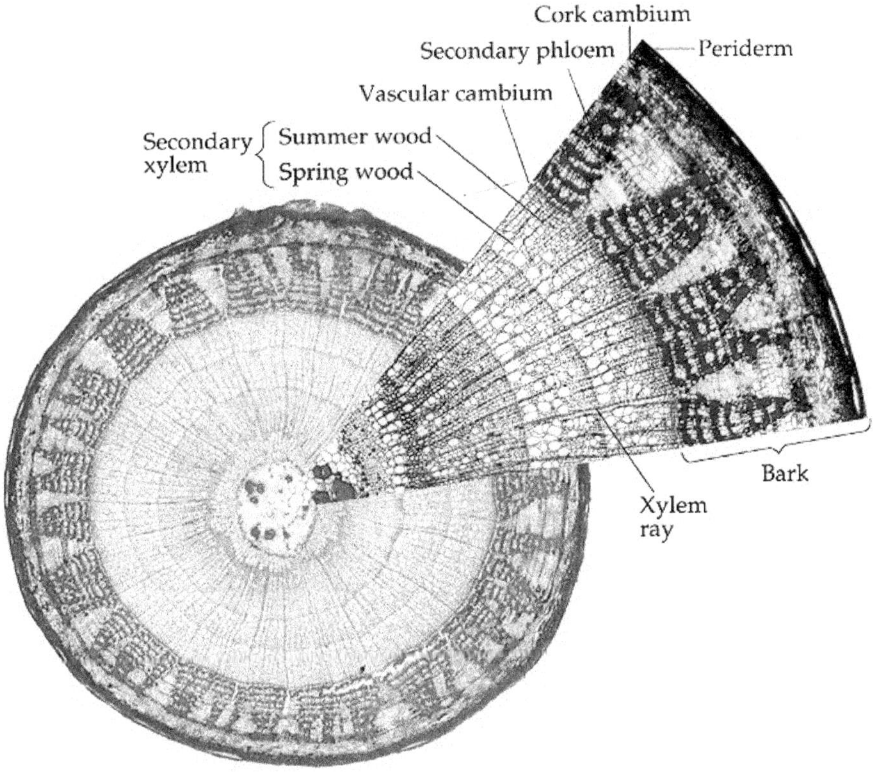

Mature woody stems, or trunks, have no epidermis or cortex, replaced by bark and wood.

Leaf organization

Leaves are the plant's photosynthetic organs with a flattened blade and a petiole attaching the leaf blade to the stem. They can be categorized as either simple or compound leaves.

Compound leaves are divided into smaller leaflets; each leaflet may have its stalk.

Simple leaves have a shape that is not deeply lobed or divided into smaller leaflets.

Leaf veins are vascular bundles within the leaves which transport water and nutrients. The top and bottom sides of the leaves are covered by the epidermis and a *waxy cuticle* that helps keep the leaf from drying out.

Along with stomata, the leaf may bear protective hairs or glands secreting irritating substances.

Modified tissues

Mesophyll tissue is a leaf's inner body, and most *photosynthesis* occurs here.

Palisade mesophyll is photosynthetic tissue with elongated *chlorenchyma cells* and many chloroplasts. Palisade is on the *upper half* of the leaf, where *light intensity* is the greatest.

Spongy mesophyll contains loosely packed parenchyma cells that increase the area for gas exchange and are in the lower half of the leaf.

As with roots and stems, leaves can be highly specialized for specific purposes.

Plants adapt their leaves to the environment; shade plants have broad leaves to maximize sunlight collection, while desert plants have reduced leaves and sunken stomata to minimize water loss. Leaves may also be large and fleshy, as in the case of succulents, to store water.

Leaves may also be protective; cactus spines are modified leaves. They can even be predatory organs, like carnivorous plants, that trap and digest insects within their leaves.

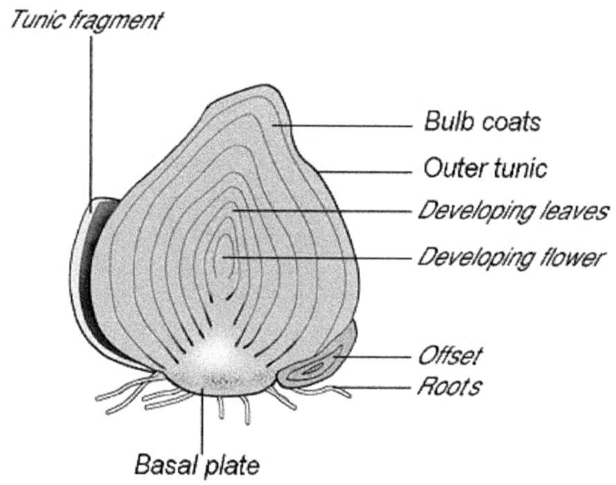

Bulb for food storage

Bulbs, such as onions, shallots, and tulips, have many layers of leaves surrounding a short stem.

Like tubers, bulbs serve as food storage.

Copyright © Sterling Education

Plant Nutrition

Essential nutrients

Essential nutrients (carbon, hydrogen, oxygen, minerals) comprise 96% of plant's dry weight.

Carbon dioxide (CO_2) is the source of carbon for a plant, water is the source of hydrogen, and oxygen can come from either atmospheric oxygen, carbon dioxide, or water.

Minerals are extracted from soil, often with assistance from fungi and bacteria.

Essential nutrients fulfill the following criteria:

1. They have an identifiable nutritional role.
2. No other element can substitute and fulfill the same role.
3. A deficiency of the element causes the plant to die.

Elements are macronutrients or micronutrients based on their concentration in plant tissue.

Macronutrients (nitrogen, potassium, and calcium) still comprise less than 5% of the plant but are present in appreciable amounts.

Micronutrients include metals (zinc and copper) in hundreds or thousands of % of the total.

Beneficial nutrients improve the growth of a particular plant but are not strictly necessary.

For example, sugar beets grow better in the presence of sodium but can survive without it.

Soil formation

Soil forms by rock weathering, freezing, glacier flow, stream flow, and chemicals.

Lichens and mosses are the first species to colonize barren rock. As they die, they deposit decaying tissues, which accumulate and support other life. Decayed organic matter is *humus.*

Humus (an organic component of soil) takes time to accumulate, but once present, its acidity leaches minerals from rocks and further enriches the soil. It keeps the soil loose and crumbly, decreases runoff, and aerates the soil.

Depending on parent material and weathering, a centimeter of soil may develop within 15 years.

Mature soil consists of soil particles, decaying organic matter, living organisms, air, and water.

A mixture of 10 to 20% humus and a top layer of soil particles is optimal for plants.

Any duplication (copies, uploads, PDFs) is illegal.

Soil composition

Soil particles are classified along a continuum, with *clay* less than 0.002 mm in diameter and the largest *sand* particles up to 2.0 mm in diameter.

Particles that are larger than clay but smaller than sand are *silt.*

Productive soil includes particles of varying sizes which provide critical air spaces.

Exclusively sandy soils lose water too readily, while clay holds water but packs too tightly.

The nitrogen content of clay soil is low because clay particles are negatively charged and cannot retain negatively charged nitrate (NO_3^-).

However, their charge does allow them to attract positively charged ions such as calcium (Ca^{2+}) and potassium (K^+).

Loam, a mixture of roughly equal sand, silt, and clay, retains water and nutrients while allowing roots to take up oxygen in the air spaces.

Animals are essential as burrowing mammals, insects, and other organisms help turn the soil.

Detrivores, organisms that feed on decomposing plants and animals, break this matter into absorbable nutrients. Fungi, protozoa, algae, and bacteria complete the decomposition process.

Decomposition is fundamental because it converts the nitrogen in dead organisms into nitrate, a form that plants utilize.

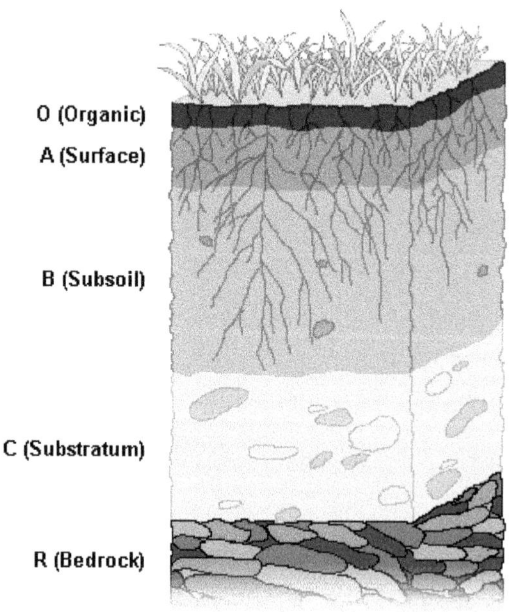

Soil stratification with organic, surface, subsoil, substratum, and bedrock layers

Copyright © Sterling Education

Transport in Plants

Water uptake

Water from soil moves directly into the cytoplasm of root hair epidermal cells and is transported across the cortex and endodermis of a root.

Symplastic movement is water and solutes passing through the continuous channel of cytoplasm that runs through plant cells via plasmodesmata.

Symplast is the area within the plant's plasma membranes through which water and solutes can travel, i.e., the connected cytoplasms of many cells.

Apoplastic movement of water and solutes through the regions outside of the symplast is collectively the *apoplast.* This includes cell walls and any extracellular spaces through which water may diffuse. It is more rapid than symplastic movement because it does not involve movement through plasma membranes, and there is less resistance to the flow of water.

Water flows through the apoplast and symplast on its way to the xylem, though mainly through the apoplast. This water contains many minerals that the plant needs but may have toxins and other undesirable substances. However, since water diffuses freely through the apoplast, there is no way to prevent the passive transport of these toxins until the water reaches the endodermis.

Casparian strip rings endodermal cells and prevents water from traveling through the apoplast; it must pass through the plasma membrane and enter the symplast to continue to the pericycle.

Plant may control how water and harmful molecules enter their vascular system. Forcing water and solutes to pass through the symplast of the endodermal cells allows the plant to regulate transport into the xylem. Once water passes under the Casparian strip in endodermal cells, it enters the apoplast again on its way to the xylem.

Mineral uptake

Minerals follow the same uptake path as water, but plant cells actively take up many minerals.

Mineral nutrient concentration in roots may be much greater than in surrounding soil. Active transport is required to uptake minerals against their concentration gradient. This may be done directly using membrane proteins that take ions into the cell.

They can also enter indirectly through a *chemiosmotic mechanism*. Proton pumps on the cell membranes of the root cells use ATP to pump H^+ ions out of the cell and into the soil. This establishes an electrochemical gradient that causes potassium, calcium, and positively-charged ions to flow into cells, as H^+ displaces cations in soil and encourages diffusion into roots.

H^+ ions associate with anions (PO_3^{4-}) in soil and bring them along when they re-enter the roots.

Any duplication (copies, uploads, PDFs) is illegal.

Nitrogen fixing bacteria

Along with the branching of roots and formation of root hairs to increase surface area, symbiotic relationships are important in nutrient acquisition. Plants cannot use atmospheric nitrogen because they lack enzymes to break the triple bond in N_2.

Rhizobia, nitrogen-fixing bacteria living in *root nodules* of legumes, makes nitrogen compounds available to plants in exchange for organic compounds.

In nitrogen fixation, rhizobia bacteria reduce atmospheric nitrogen (N_2) to ammonium (NH_4^+).

Other plants may not have nitrogen-fixing bacteria in their roots but have a relationship with free-living nitrogen-fixers. The relationship between roots and fungi is crucial.

The branching threads of the fungus, called *hyphae,* are robust nutrient absorbers due to their large surface area and ability to decompose decaying matter into inorganic nutrients.

Hyphae penetrate root tissues and deliver nutrients to the plant, which they absorb from the soil. The association between roots and hyphae is *mycorrhiza.*

A plant can uptake many more nutrients in conjunction with a fungus than it could on its own due to the increased absorption capabilities of the fungus. In return, the plant delivers carbohydrates and amino acids to the fungus.

Mycorrhizae allows plants and fungi to inhabit soils where they may otherwise die.

Not all plants benefit those from whom they gather nutrients. Some plants are parasites, sending out haustoria to grow into the host plant and tap its xylem and phloem.

In an engaging reverse of the mycorrhizal relationship, some plants parasitize fungi and steal organic compounds that they receive from other plants.

Plants can even take their nutrients from animals, as with the Venus flytrap and the sundew, which are carnivorous and prey on insects.

Mechanical transport

Transportation in plants relies on the mechanical properties of water:

1) uptake and release of water and solutes by individual cells, as when root cells absorb water and minerals from the soil

2) short-distance transport of substances from cell to cell (e.g., loading of sucrose from photosynthetic cells into the sieve tube cells of the phloem)

3) long-distance transport of sap within the xylem and phloem.

Copyright © Sterling Education

Properties of water

Diffusion and water potential rely on water's *polarity* and *hydrogen bonding* to transport water, minerals, and organic molecules.

Water potential is the potential energy of water and is a measure of the capacity to release (or take up) water. Water flows from a higher region to a lower water potential.

In cells, water potential includes both the *pressure potential*, which is the effect that pressure has on water potential, and *osmotic potential*, the effect that solutes have on water potential.

Water flows from an area of high to an area of low pressure, just as it flows from an area of low solute concentration (high water concentration) to an area of high solute concentration (low water concentration).

When the solute concentration is higher within a cell than outside, water enters and increases the pressure inside the cell.

Turgor pressure

Turgor pressure is the pressure potential that increases due to osmotic pressure. Sturdy cell walls allow substantial water pressure to build until the *osmotic potential* is balanced and water ceases to enter.

Cells maintain a *high solute concentration* within the cytoplasm to encourage water movement into the cells.

Hypotonic solution has a low solute concentration and high-water concentration.

Animal cells swell and undergo lysis if placed in a *hypotonic* environment, but plant cells *must* have hypotonic extracellular fluid to maintain the overall turgidity of the plant's body.

Transpiration

Xylem movement is a passive process created by natural properties. Water can rise upward through the xylem by pushing and pulling. These are forces, but they arise naturally rather than by active transport.

Xylem cells are non-living and do not perform any processes to move water.

When transpiration is low at night, water entering root cells creates osmotic *root pressure*. This is a positive pressure that pushes the sap upward through the plant. It can cause water to be forced out of leaf veins, a phenomenon of *guttation.*

Guttation is visible as water beads along the edges of leaves.

Any duplication (copies, uploads, PDFs) is illegal.

Root pressure

Tall trees generate almost no root pressure since the weight of the water pushing down on the xylem more than counteracts the positive pressure. However, root pressure is not usually sufficient to drive water movement.

Root pressure is less critical for most plants than *transpiration pull*, which is primarily responsible for upward sap movement in the daytime.

Transpiration pull is created by transpiration, the evaporation of water from leaves. This drives negative pressure, or tension, which pulls the sap upward.

Cohesion-tension model

While the mechanisms behind upward movement through the xylem are not entirely understood, the most generally accepted theory is the *cohesion-tension model.*

Under this model, the cohesive properties of water and the tension created by transpiration pull are the primary forces driving sap against gravity.

Cohesion refers to water bonding together due to hydrogen bonds. The cohesion of water is strong enough to transmit the pulling force down to the roots.

Adhesion is the ability of water to cling to other molecules, such as the walls of xylem vessels. It gives a water column extra strength to resist gravity and prevent slipping back down.

Transpiration pull, cohesion, adhesion, and the structure of xylem vessels are all important in the carrying of water by the transpiration stream.

Transpiration stream is an unbroken column of water that flows from the roots to the leaves and then evaporates into the air. It must remain continuous to conduct water upwards properly.

Tension created by the transpiration of water coupled with the cohesive and adhesive forces is enough to support the stream against the forces of gravity, even in trees 100 meters tall.

Plants must constantly balance the requirements of the transpiration stream against the water loss incurred by the process of transpiration. Losing too much water through the leaves forces the plant to close its stomata, inhibiting gas exchange, and photosynthesis.

Copyright © Sterling Education

Abiotic factors driving transpiration

Guard cells regulate the opening and closing of stomata and respond to four main abiotic factors that affect the rate of transpiration in a typical terrestrial plant:

1. *Light* – the transpiration rate is much greater during daylight hours since photosynthesis occurs only when light is available. To maximize photosynthesis, more stomata remain open, and more water is transpired.

2. *Humidity* – water diffuses from a leaf from a high concentration gradient inside the leaf to a lower concentration gradient in the air. As leaves transpire, their water vapor remains close to the leaf.

   Humidity can increase around the leaf and inhibit further transpiration. A rise in humidity means a higher water vapor concentration outside the leaf, decreasing the transpiration rate.

3. *Temperature* – a rise in temperature excites water molecules and increases the transpiration rate. It can cause the photosynthesis enzymes to function efficiently, increasing the transpiration rate to sustain the increased rate of photosynthesis.

4. *Wind* – can offset the effect of humidity by removing the humidity from around the leaf, thereby increasing the transpiration rate.

Phloem tissue transport

Phloem tissue transports sugars and organic molecules from sources to sinks. During the growing season, leaves photosynthesize vigorously and produce massive quantities of sugar.

Phloem transport does not rely on transpiration pull, so it is only indirectly affected by changes in the four abiotic transpiration factors

Sources are in the leaves and storage organs, while *sinks* are common in flowers, seeds, roots, or other parts of the plant that require nutrients.

Between the sinks and the sources are the sieve tubes, formed by sieve-tube cells arranged end to end. Although it seems counterintuitive, phloem transport requires energy despite it often moving in the direction of gravity.

Loading is the movement of source sugars into the phloem, which goes against the concentration gradient.

Active translocation describes this energy-requiring phloem transport.

Pressure-flow model outlines how a positive pressure potential drives the direction of phloem transport. It proposes that sucrose is loaded into the phloem via active transport due to an electrochemical gradient established by an H^+ pump.

Any duplication (copies, uploads, PDFs) is illegal.

Phloem loading results in a high solute concentration at the source end. This creates high osmotic potential in the phloem, encouraging water to diffuse into the phloem.

The buildup of pressure at the source creates a positive pressure potential, which drives the sap to flow to a region of lower pressure – the sink. At the sink, the sap is unloaded down its concentration gradient, exiting the phloem, and entering the tissue for which it is destined.

Water is recycled via transport in the xylem.

Contrary to widely held belief, phloem sap does not necessarily move from the leaves to roots. It merely flows from source to sink, meaning it can move in any direction along the phloem.

For example, the source may be at an underground tuber and the sink at a fruit; in this case, sap flows upwards against the direction of gravity.

The common mnemonic for phloem and xylem transport, "*phlo low, xy high*," is misleading.

While xylem transport always moves "*high*," phloem transport can move in either direction.

Copyright © Sterling Education

Control of Growth and Responses

Responding to stimuli

One defining characteristic of life is the ability to respond to stimuli. Adaptive organisms respond to environmental stimuli because they lead to the longevity and survival of the species.

Plants lack nerves and muscles and must respond by varying their growth patterns.

Tropisms are directional movement responses that occur due to external environmental stimuli. Hormones respond to the stimulus by directing one side of the plant to elongate, resulting in a curve toward or away from a stimulus. The stimulus comes from only one direction and affects the direction of growth.

Growth toward a stimulus is *positive tropism*, while growth away is *negative tropism*.

Phototropism is the growth of plants in response to light, commonly seen in shoots and roots. This response is controlled by a class of plant hormones called *auxins.*

Shoots display *positive phototropism* when auxins produced at the tip of the shoot are translocated to the side of the shoot tip that receives *less light*.

Once auxins reach the shaded side of the plant, they initiate faster cell growth on only that side. Plant shoots gradually bend toward the light as the shaded side elongates faster than the sunny side. Some plants do this quickly, following the sun as it moves across the sky daily.

Negative phototropism is observed in roots, where auxins cause them to grow away from light and into the soil when auxins initiate elongation on the upper surface of the roots, bending the roots *downward.*

Gravitropism

Auxins similarly affect plant organs for gravity. *Gravitropism* is a response to the direction of gravity. Stems and leaves exhibit negative gravitropism when they grow upward in opposition to gravity. This can be seen in an upright plant on its side, projecting its leaves upward.

Phototropism is observed as the leaves re-orient themselves to best face the light.

Positive gravitropism aids phototropism in encouraging roots to grow downward. Root cap cells detect gravity using *statoliths*, starch grains within starch *amyloplasts* (storage organelles). Wherever the statoliths settle due to gravity, the root tip grows in that direction.

Ethylene hormone works with auxins in the *thigmotropism* response.

Thigmotropism response is a growth response due to contact with solid objects. A few minutes of contact can bring about a response lasting several days.

Any duplication (copies, uploads, PDFs) is illegal.

Cells in contact with the object are directed to grow less, while those on the opposite side elongate, causing the tendril to bend around the post. Combined with phototropism and gravitropism, the tendril spirals upward. For example, the coiling of morning glory or pea tendrils around posts is a typical example of positive thigmotropism.

Roots exhibit negative thigmotropism when growing away from a soil obstacle, such as a rock.

In contrast to tropisms, *nastic movements* are independent of the direction of the stimulus.

Seismonastic movements

For example, *seismonastic movements* result from touching, shaking, or thermal stimulation. The leaflets of *Mimosa pudica* exhibit a seismonastic movement when they curl up when touched. This response takes only a second or two due to a loss of turgor pressure within cells, causing the petiole to droop.

Another example is the Venus flytrap, which snaps down on an insect when it touches a trigger hair at the base of the trap.

Circadian rhythms

Sleep movements are nastic responses to the daily changes in light level, as with the prayer plant that folds its leaves each night. This is an example of a *circadian rhythm*, a biological response on a 24-hour cycle.

Circadian rhythms are driven by a *biological clock*, an internal mechanism that roughly maintains the rhythm without stimuli.

With the proper external stimuli (e.g., light from the sun), the biological clock synchronizes to 24 hours. It detects sunlight using a red flavin pigment *cytochrome*, which absorbs the blue wavelength of light.

Auxins

For plants to respond to stimuli, the activities of plant tissues must be coordinated. Because plants do not have a nervous system, communication is done through hormones.

Hormones are chemical messengers synthesized in one part of a plant after receiving a stimulus. They may travel long distances to another part of the organism via the phloem or move from cell to cell via carriers. Several hormones may influence response and require a specific ratio of multiple hormones.

Synthesized hormone imitators are *plant growth regulators* (PGRs) manipulating plant growth.

Auxins are vitally crucial to the processes of growth and development. They drive *apical dominance*, inhibiting *lateral bud growth* while promoting the *shoot tip's growth.*

 Copyright © Sterling Education

For example, auxins direct fruit growth, prevent leaf abscission (falling) and senescence (aging or dormancy), and help initiate flowering.

Most auxins are produced in *shoot apical meristem*, traveling down the plant to their destination. They regulate growth by initiating cell elongation via poorly understood mechanisms.

Cytokinins

Cytokinins are a class of plant hormones that promote cell division rather than elongation.

Cytokinins enhance auxin effects when they prevent leaves' senescence.

Cytokinins oppose auxin effects by promoting abscission and inducing lateral growth.

Auxins to *cytokinins ratio* determines which process prevails.

Gibberellins are another group of 70 chemically similar plant hormones.

Like auxins, gibberellins promote growth by cell elongation and other mechanisms.

Gibberellin GA_3 releases sugars in seeds to feed the embryo and facilitate germination.

Abscisic acid

Despite its name, *abscisic acid* (ABA) rarely plays a role in leaf abscission. Although ABA promotes abscission when applied by researchers, the hormone does not appear to have this function in the plant naturally. Instead, ABA competes with gibberellins, promoting *dormancy* instead of germination and growth.

As for auxins and cytokinins, the ABA to gibberellins *ratio* determines the prevailing hormone.

Reductions in ABA and increases in gibberellins break seed and bud dormancy, causing seeds to germinate and buds to produce leaves.

ABA responds to water deficiency by closing stomata.

High ABA concentrations keep a seed or bud dormant, stopping all growth. ABA initiates dormancy in plant organs, usually in preparation for winter. It moves from leaves to vegetative buds in the fall, converting them to *winter buds* covered by thick, hardened scales.

Any duplication (copies, uploads, PDFs) is illegal.

Ethylene

Ethylene is a gaseous hormone responsible for positive feedback loops such as ripening, abscission, flowering, and senescence.

Ethylene gas ripens fruit by increasing *cellulase* production, an enzyme that hydrolyzes plant wall cellulose and softens fruit. As the fruit ripens, it releases additional ethylene gas, stimulating further ripening.

At various stages of a plant's life cycle, ethylene may promote or inhibit certain growth events. The hormone easily diffuses out of cells and into the atmosphere, which is why a ripening apple can induce the ripening of a banana some distance away.

Ethylene stimulates *cellulase production* to initiate the abscission of leaves or fruit, provided there are sufficiently low levels of auxin. Cellulase weakens the connections of the leaves or fruits to the stem and causes them to drop.

Researchers cannot agree on why plants shed leaves in the fall, but it may be for plants to conserve energy for other activities during wintertime, such as developing seeds in preparation for spring. Fruits abscesses become more available for animals to eat and transfer the seeds.

Photoperiodism

Photoperiodism is a physiological response to relative lengths of daylight and darkness and is important for flowering since changes in day and night length signal the onset of spring.

Many physiological changes in plants (e.g., seed germination, breaking of bud dormancy, and the onset of senescence) are related to seasonal changes in day length.

Plants are divided into three groups based on *photoperiodic flowering* strategy.

Short-day plants flower when the day is shorter than a critical length or when the night is longer than a critical length. After discovering photoperiodism, researchers understood that it depends on the length of the night, not the length of the day. Short-day plants may more accurately be termed *long-night plants*.

Long-day plants (or *short-night plants*) flower when the day is long; thus, the night is shorter than a critical length, often during the summer.

Day-neutral plants are unaffected by day length and rely on external or internal cues to flower.

In addition to cytochrome for blue light, plants have *phytochrome*, a blue-green leaf pigment in two forms. P_r and Pfr can be converted interchangeably.

Pr (or phytochrome red) absorbs red light (660 nm).

Pfr (or phytochrome far-red) absorbs far-red light (730 nm).

 Copyright © Sterling Education

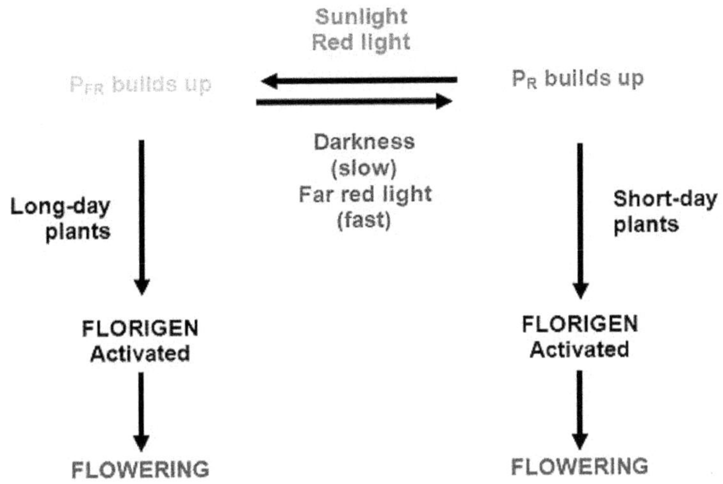

Phytochrome is a blue-green leaf pigment in P_r and Pfr forms

During 24 hours, there is a shift in the ratio of pigments.

Direct sunlight has more red than far-red light, so more Pr is converted into Pfr during the day.

Shades and sunsets have farther-red than red light, so Pfr is converted to Pr at sunset and night.

P_{fr} remaining when night ends are the way for plants to time the length of dark periods.

Short nights result in higher P_{fr} levels in mornings, while a long night depletes most of it.

P_{fr} promotes flowering in long-day (short-night) plants, so a short night induces flowering.

P_r promotes flowering in short-day (long-night) plants, and P_{fr} inhibits it, so they flower after a long night when P_r is high, and P_{fr} is low.

Any duplication (copies, uploads, PDFs) is illegal.

Notes for active learning

 Copyright © Sterling Education

Sporophyte and Gametophyte Generations

Alternation of generations

Plants have a two-generation life cycle with an alternation of generations, which includes the sporophyte and gametophyte generations.

Gametophyte generation dominates nonvascular terrestrial plants (liverworts, hornworts, and mosses). Gametophyte represents most of the plant's body, while the sporophyte is a small, protruding structure that relies on the gametophyte. Once it grows from a spore, the gametophyte develops organs in which mitosis produces the gametes.

Archegonia (singular, *archegonium*) produces eggs, while *antheridia* (singular, *antheridium*) produce sperm. Like in animals, these sperms have flagella. Gametophytes must be in a damp environment so that its sperm can swim through water from the antheridium to an egg in an archegonium.

Sperm of one gametophyte may travel to the egg of a different gametophyte, or the gametophyte may be self-fertilizing. Once sperm reaches the archegonium, it fertilizes the egg to form a zygote which divides via mitosis, developing into a small sporophyte growing on the larger gametophyte.

Sporophyte is composed of a *foot*, which arises from the former archegonium and anchors it to the gametophyte, and a *capsule* at the end of a short stalk.

In contrast to bryophytes, the dominant generation in most vascular plants is the sporophyte. However, note that many seedless vascular plants have gametophytes that are not dependent on sporophytes.

Sporangium

Sporangium is a spore-producing tissue that continually undergoes meiosis to form the haploid spores stored in the capsule that splits to release spores, which travel on the wind or through the water from the original plant. Once there, a *germ tube* emerges from the spore and divides into the gametophyte.

Rhizoids anchor the gametophyte to the soil and absorb water and minerals. The gametophyte projects antheridia and archegonia, which produce sperm and eggs to begin the cycle again.

All three phyla of nonvascular terrestrial plants roughly follow this life cycle, with deviations.

In hornworts (phylum *Anthocerotophyta*), the archegonia are recessed into the plant's body.

Liverworts and mosses have long-necked, hollow archegonia down which the sperm must swim to reach the egg.

Any duplication (copies, uploads, PDFs) is illegal.

Fern life cycle begins with the production of spores by meiotic cell division within sporangia located in clusters on the underside of the fronds. Spores are released and dispersed by wind.

Each spore germinates into a *prothallus*, which is an immature gametophyte. The prothallus develops antheridia and archegonia on its underside and begins generating gametes.

Fertilization

Fertilization occurs only if water is present to allow flagellated sperm to swim from antheridia to an archegonium. The resulting zygote develops inside an archegonium, but the embryo quickly outgrows the space. A sporophyte becomes visible when the first leaf grows above the prothallus, and roots develop below. Fronds are projected from the roots, eventually maturing into a large, leafy sporophyte.

Because seed plants are heterosporous, they have two distinct types of sporangia. Microsporangia produce microspores via meiosis. Each microspore divides mitotically into a microgametophyte. The gametophyte generates two sperm cells from its antheridia. While the gametophytes of the seedless vascular plants are small but still observable, those of the seed plants are tiny and entirely dependent on the sporophyte.

Ovules house megasporangia, producing the megaspores that divide into a megagametophyte with an archegonium that produces the eggs.

Gymnosperms

Gymnosperm life cycle begins with the dominant sporophyte (the entire tree) and its sporangia borne in cones. Each scale-like sporophyll of a pollen cone, called a *microsporophyll,* has two microsporangia. Microspores generated in the pollen cones develop into the male gametophyte, a small cluster of cells known as a *pollen grain*.

Ovule (seed cones) has *megasporophylls*, with *megasporangia* generating *megaspores*.

Megaspores are much larger than male microspores. Only one megaspore develops into a megagametophyte with 2 to 6 archegonia, each containing a large egg.

Pollen grains make their way to ovules of seed cones and develop a pollen tube that digests its way toward a megagametophyte, discharging two non-flagellated sperm.

Fertilization takes place one year after pollination.

Ovule matures and becomes a *seed*, composed of the *embryo*, *reserve food*, and *seed coat*.

Woody seed cones open to release winged seeds in the fall of the second season.

Some gymnosperms produce fleshy *arils* that are edible by animals and dispersed similarly to angiosperm fruits.

Copyright © Sterling Education

Angiosperms

Angiosperm life cycle is similar, differing in anatomy.

Angiosperm sporangia and *gametophytes* are in flowers rather than cones.

Flowers are reproductive structures produced by the *shoot apical meristem.*

Flowers have petals, sepals, and female and male reproductive structures as *complete flowers.* These are *bisexual* with male and female reproductive organs; *hermaphroditic.*

Many angiosperms have *incomplete flowers,* missing one or more components.

Flower that is only male is *staminate.*

Flower that is only female is *carpellate.*

Plant species may be *dioecious,* in which there are designated male plants with staminate flowers and designated female plants with carpellate flowers.

Plants may be *monoecious*, each with staminate and carpellate flowers and self-fertilizing.

This differs from a hermaphroditic species since the flowers are unisex, not bisexual.

Carpellate flowers

Carpellate flowers are named for their *carpels*, the female reproductive structures.

A carpel is a modified sporophyll located at the center of a flower.

Pistil is the female reproductive structure of a flower; it may refer to a single carpel or several fused carpels. The vase-like carpel is generally composed of three parts.

Within the enlarged base of a carpel is the *ovary.* A slender stalk of the *style* extends from the ovary. At its tip is the *stigma,* an enlarged sticky knob that receives pollen grains.

An ovary contains ovules, making the carpel analogous to the seed cone of a gymnosperm.

Each ovule is covered by parenchymal cells except for one small *micropyle* opening.

One parenchyma cell enlarges to become a *megasporocyte* (sporocytes being the cells that comprise sporangia), which generates four haploid megaspores.

Three megaspores are nonfunctional, but one divides mitotically into a seven-celled megagametophyte as an *embryo sac.*

Five cells are support cells, two of which are *synergids,* and three are *antipodal.* The other two cells are the egg and binucleate cells, with two nuclei as *polar nuclei.*

Any duplication (copies, uploads, PDFs) is illegal.

Staminate flowers

Staminate flowers are named for their *stamens,* which are the male reproductive structures.

Stamen filament is a slender stalk that supports a saclike container at the top: the *anther.*

The stamen is analogous to the pollen cone of a gymnosperm; it has microsporangia within the anther that generate microspores, which develop into microgametes that produce pollen grains.

Anthers have four pollen sacs, each containing many *microsporocytes.*

Microsporocytes undergo meiotic cell division to produce four haploid microspores; each divide into two haploid cells enclosed in a sculptured wall. This is a pollen grain containing a *tube cell* and a *generative cell.*

The larger tube cell eventually forms the pollen tube, while each generative cell divides mitotically to form two sperm. Once both events have occurred, the pollen grain has become the mature male gametophyte.

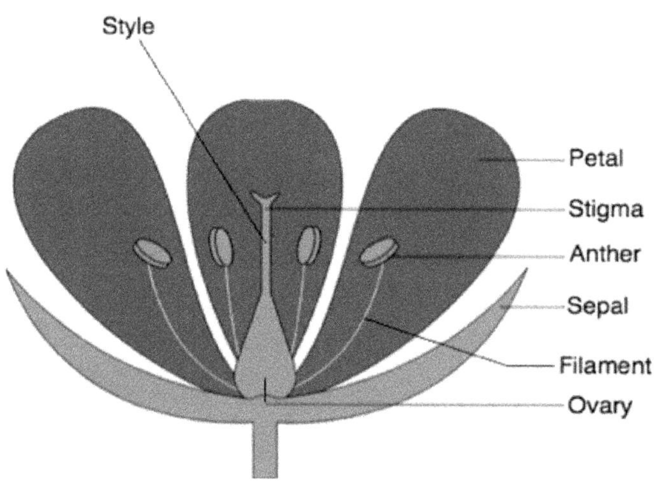

Dicotyledonous animal-pollinated flower

Copyright © Sterling Education

Reproduction in Plants

Pollination

Pollination occurs when pollen is transferred from an anther to a stigma. It begins the breakdown of walls around the pollen sacs in the anther.

Wind or animals carry pollen to the flower's stigma, where it fully matures into a pollen grain.

Self-pollination is the transfer of pollen from the anther to a stigma on the same plant.

Cross-pollination is the transfer of pollen from the anther to the stigma of another plant.

Plants often have mechanisms that promote cross-pollination, such as the carpel only maturing after anthers have released their pollen.

Cross-pollination is advantageous because it *increases genetic diversity.*

Once a pollen grain reaches a stigma, it must travel down the style into the ovary. This is accomplished by the tube cell, which grows into a pollen tube that extends through the stigma and style to the micropyle opening of an ovule.

Pollen tube injects its two sperm cells into the embryo sac so fertilization can occur.

Double fertilization in angiosperms

Unlike many other organisms, angiosperms utilize *double fertilization*, with two fertilization events. One of the sperm cells in the pollen grain fertilizes an egg to form a diploid zygote, while the other unites with the two polar nuclei to form a triploid cell as the *endosperm.*

Ovule develops into a seed with three main components:

zygote divides mitotically to form the embryo

endosperm nucleus divides into multicellular tissue that nourishes the embryo

walls of the ovule develop into a protective seed coat or *integument.*

Angiosperms conserve energy by only investing in endosperm production if an egg is fertilized. Without double fertilization, they would have to produce nutritive tissues in anticipation of fertilization. As with menstruating mammals, this tissue goes to waste if the egg is not fertilized.

Additionally, double fertilization is an advantage because it speeds the development of the endosperm, with three nuclei rather than only one.

Any duplication (copies, uploads, PDFs) is illegal.

Single fertilization in gymnosperms

Note that gymnosperms do not undergo double fertilization. Only one of the sperm cells fertilizes the egg, while the other recedes.

Embryo is fed by nutrients passed along by its cotyledons from the tissue of the mother cone.

Pollen falls on stigma → pollen germinates → pollen tube grows → two sperm enter ovule → egg is fertilized and becomes embryo → polar bodies are fertilized and become endosperm → the entire ovule becomes a seed and the ovary becomes a fruit.

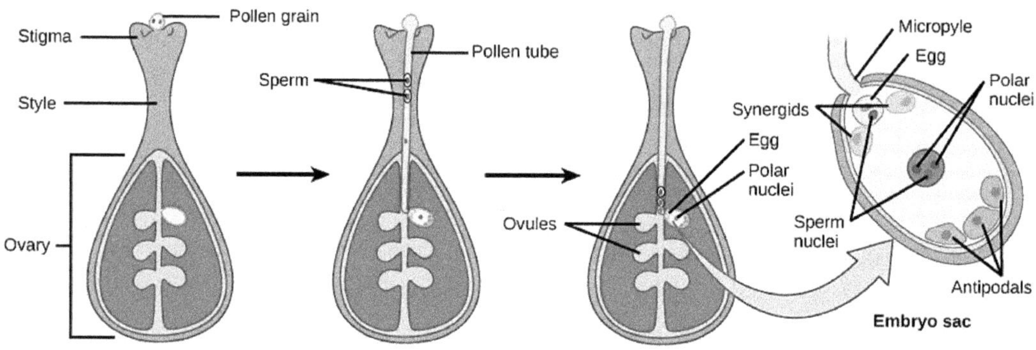

1	2	3	4
Pollen grain adheres to the stigma, which contains two cells: a generative cell and a tube cell.	Pollen tube cell grows into the style. The generative cell travels inside the pollen tube. It divides to form two sperm.	Pollen tube penetrates an opening in the ovule called a micropyle.	One sperm fertilizes the egg to form the diploid zygote. The other sperm fertilizes two polar nuclei to form the triploid endosperm, which will become a food source for growing embryo.

Copyright © Sterling Education

Seeds and Fruits

Seed development

After double fertilization, the single-celled zygote of an angiosperm divides asymmetrically to form the proembryo and the suspensor.

Proembryo is an undifferentiated ball of cells which the *suspensor*, analogous to an umbilical cord, anchors to the endosperm. At this stage, the embryo is globular.

Embryos begin to have a defined symmetry as cells near the suspensor differentiate into root cells, and those at the opposite end become shoot cells to establish the *root-shoot axis.*

After differentiation into the proembryo and suspensor, one or two cotyledons develop, depending on whether the plant is a monocot or eudicot.

Cotyledon of a monocot embryo rarely stores food. Instead, it absorbs food molecules from the endosperm and passes them to the embryo.

Conversely, the eudicot embryo has two cotyledons, which usually store the embryo nutrients. The cotyledons are noticeable in a eudicot embryo and may fold over.

Whether monocot or eudicot, embryo has a heart shape when the cotyledons appear but then grows to a torpedo shape.

With elongation, root and shoot apical meristem become distinguishable at the ends of the plant.

Embryo differentiation

Embryo continues to differentiate into three regions.

Epicotyl is embryonic shoot tip above cotyledons, contributing to shoot and leaf development.

Hypocotyl is below the cotyledons is the embryonic stem.

Radicle is the embryonic root at the bottom.

Any duplication (copies, uploads, PDFs) is illegal.

Dormancy

Seed germination requires certain conditions; otherwise, the seed remains dormant, undergoing sufficient respiration and metabolism to keep the seed alive but at a low level.

Seeds retain viability for varying times: maple seeds last a week, while lotus seeds are viable for hundreds of years.

Dormancy allows the seed to survive adverse conditions and ensure germination begins during optimum conditions. One such condition is sufficient oxygen so the plant can undergo aerobic cell respiration to provide energy until the first leaves emerge. Once the leaves emerge, photosynthesis can provide the energy needed for growth if there is sufficient light.

Water also causes seeds to swell and burst out of the seed coat, enabling the seedling to emerge. Seeds do not germinate if water is needed to rehydrate the cells of the seed and activate certain enzymes that initiate its metabolism.

Appropriate temperatures are needed for enzymes. Enzyme activity is too low for germination if it is too hot or cold.

Seed germination

Once germination of a eudicot or gymnosperm embryo begins, the radicle emerges and burrows into the soil, becoming the taproot. The epicotyl then begins to grow upward through the soil.

Germination proceeds differently depending on the species. In some instances, the cotyledons travel with the epicotyl aboveground, while in others, they remain underground.

If the cotyledons are *above ground*, young leaves sprout from the tip of the epicotyl (now a rudimentary shoot, the *plumule*); cotyledons may die or persist and become photosynthetic.

If the cotyledons are *below ground*, they may die or convert into storage organs.

In monocots, the plumule and radicle are enclosed in protective sheaths as the *coleoptile* and the *coleorhiza*, respectively. The radicle bursts from the coleorhiza upon germination, while the plumule escapes the coleoptile when it reaches the soil's surface, upon which the first true leaves burst through their casing.

Copyright © Sterling Education

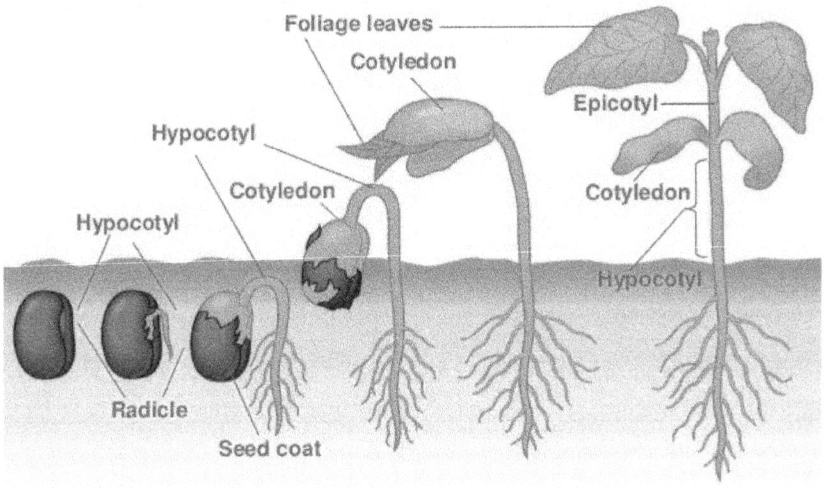

Seed germination from radicle (the embryonic root of the plant) through hypocotyl, cotyledon, and flowering leaves

Fruit types

Fruit is a mature ovary that encloses seeds to protect and facilitate seed dispersal. It develops as the ovary wall ripens and thickens into a protective and nutritious covering as a *pericarp.*

Flower parts may become incorporated into the fruit. Often, the flowers, rather than the ovary, are the bulk of the fruit and are an *accessory fruit.*

There are two types of fruit:

1) *simple fruits* and

2) *compound fruits.*

Simple fruits develop from a single carpel, made of one or several fused ovaries. Legumes and cereal grains are simple *dry fruits*; often mistaken for seeds because they have a dry pericarp tightly adhering to the seed.

Simple fruits have fleshy pericarp (e.g., tomatoes, plums, and peaches).

Compound fruit develops from several individual ovaries. If the ovaries are present in a single flower, it is an *aggregate fruit* (e.g., a blackberry). Several flowers may ripen together to form *multiple fruits* (e.g., a pineapple).

Aggregate fruit develops from many ovaries on a *single flower.*

Multiple fruits develop from many flowers on a *single plant.*

In both cases, fruit is made of many different fused fruits.

Any duplication (copies, uploads, PDFs) is illegal.

Seed dispersal

The goal of a fruit is to facilitate seed dispersal. Dispersal moves seeds some distance from the original plant, expanding the species' range, incorporating greater genetic diversity, and reducing intraspecies competition.

Fruits accomplish this in various ways, using other organisms or the plant's agency.

For example, plants such as clover, bur, and cocklebur have hooks and spines, which attach to the fur of animals.

Other plants have lightweight hairs, plumes, or wings that disperse in the wind. These fruits do not resemble the familiar apples or bananas; they are inedible to humans and other animals.

Many fruits are dispersed by water or gravity as they fall to the ground. Some plants can forcefully eject their seeds from the plant rather than relying on gravity alone.

Plants utilize other organisms, especially birds and mammals, by making their fruits enticing.

Edible fruits are eaten by birds and mammals and defecated some distance away. In some cases, they may be buried by rodents like squirrels.

Fragmentation and vegetative propagation

Many plants are capable of asexual reproduction, which can confer significant advantages.

Many algae and bryophytes use asexual reproduction by *fragmentation* when they detach *gemmae cups,* single cells, or masses of cells that develop into new individuals.

The vascular plant version of fragmentation is *vegetative propagation,* which utilizes the meristematic tissue of a parent plant to produce clonal offspring. For example, the eyes on a potato plant tuber undergo vegetative propagation when they bud to produce new plants.

Stolons and rhizomes are another way in which plants facilitate *vegetative propagation.* This strategy can greatly benefit because it is rapid, less energy-consuming, and less reliant on pollinators than sexual reproduction. It allows the species to expand its range and ensure survival by creating a *clonal colony.*

A plant on its own may die in harsh conditions and be unable to regrow, meaning its genes fail to propagate. However, a colony of identical clones dramatically increases the chance that at least one plant survives and continues to propagate the colony's genome.

Vegetative propagation can have adverse side effects because it rapidly multiplies the effects of the disease. Plants are vulnerable to disease because cells are connected via *plasmodesmata,* creating a superhighway for pathogens throughout the entire organism.

Vegetative propagation enhances the likelihood that the parent plant spreads any pathogens to the entire clonal colony, leading to a mass die-off.

Copyright © Sterling Education

Apomixis

Some plants have developed a workaround to this downside, called *apomixis,* a form of asexual reproduction that involves seeds. Rather than propagating itself by sprouting clones from its body, the plant disperses seeds as it would in sexual reproduction.

However, the seeds are not formed from a combination of sperm and eggs but are generated only from the mother. Typically, once the mother generates a megaspore within an ovule, it undergoes meiosis to generate four haploid spores.

During apomixis, however, the megaspore divides by mitosis instead to form two diploid spores. These spores develop into gametophytes capable of autonomously generating an embryo without fertilization from a father plant.

Asexual reproduction by apomixis lowers the risk of disease found in vegetative propagation but has the disadvantage of decreased genetic diversity and a slow evolution.

Apomixis is of interest to agricultural scientists who wish to engineer sexually reproductive crops to reproduce via apomixis instead, making crop production efficient. Most interferences in plant reproduction have historically focused on breeding or vegetative propagation.

Hybridization

Humans have generated plants for thousands of years by planting cuttings from the parent plant. Advances in tissue culture and manipulation of plant hormones have resulted in new methods for producing commercially valuable plants.

Hybridization has become highly sophisticated. Science is capable of genetic engineering, in which genes are altered to create *transgenic plants* with insects, pathogens, herbicides, drought, or frost resistance traits.

Scientists are attempting to introduce C_4 photosynthesis and improve the efficiency of water and CO_2 uptake in crop plants to improve agriculture. They also wish to induce plants to produce chemicals with industrial applications.

Genetic engineering has medical applications: researchers have designed plants that produce specific antibodies that are promising treatments for cancer and other diseases.

As molecular biology and genetics become more advanced, the possibilities continue to grow.

Any duplication (copies, uploads, PDFs) is illegal.

Notes for active learning

Copyright © Sterling Education

Notes for active learning

Notes for active learning

 Copyright © Sterling Education

REVIEW

Photosynthesis

Photosynthetic Organisms

Physical and Chemical Aspects of Photosynthesis

Photosynthetic Structures

Photosynthetic Reactions

Noncyclic Electron Pathway

Cyclic Electron Pathway

Calvin Cycle

Types of Photosynthesis

Page intentionally left blank

Copyright © Sterling Education

Photosynthetic Organisms

Converting solar energy to chemical energy

Photosynthesis converts solar to chemical energy which drives the assembly (*anabolism*) of organic molecules and fuels cellular metabolism.

Plant leaves, composed of the *lamina* (blade) and the *petiole* (stalk), are the plant's primary photosynthetic organs and typically have a large surface area to maximize light harvesting.

Simple leaf has one *lamina*, while a *compound leaf* has many distinct laminae.

Leaves are highly variable in shape and usually thin, so light penetrates cells on the underside.

Photosynthesis includes:

1) *light-dependent reactions* when solar energy is captured and

2) *Calvin cycle* (light-independent or *dark* reactions) making carbohydrates (sugars).

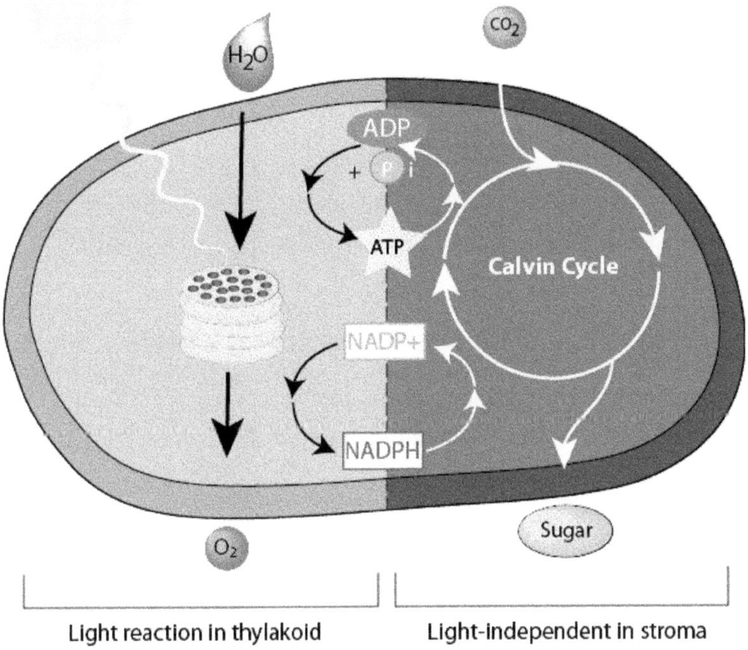

Light-dependent reactions produce ATP and NADPH, light-independent produce sugars

Any duplication (copies, uploads, PDFs) is illegal.

Photosynthesis

Photosynthesis is a *backbone energy process* crucial for life. Only plants, algae, and certain bacteria perform photosynthesis, but all organisms undergo cellular respiration (i.e., the breakdown of nutrients into energy).

Photosynthesis is a complex metabolic pathway:

light energy + carbon dioxide + water → carbohydrate + oxygen + water

light energy + 6 CO_2 + 12 H_2O → $C_6H_{12}O_6$ + 6 O_2 + 6 H_2O

Photosynthesis occurs in *chloroplasts* (plants).

Cellular respiration

Cellular respiration occurs in *mitochondria* in eukaryotes (plants and animals).

Cellular respiration breaks down (*catabolism*) carbohydrates (e.g., *glucose*) produced from photosynthesis into energy (*ATP*) to fuel metabolic activities:

carbohydrate + oxygen → carbon dioxide + water + energy

$C_6H_{12}O_6$ + 6 O_2 → 6 CO_2 + 6 H_2O + ~34 ATP

Photosynthetic autotrophs

Primary producers (or *autotrophs*) transform inorganic elements into organic compounds (e.g., glucose) created by photosynthesis to fuel *primary producers* and organisms *above them in the food chain.*

Most primary producers are *photosynthetic autotrophs,* using light energy for sugar synthesis.

Chemosynthetic autotrophs are producers; inorganic reactions form organic compounds. For example, food chains rely on photosynthesizing organisms (except for rare life based on chemosynthetic autotrophs).

Photosynthesis rates

Rate of photosynthesis measures O_2 production and CO_2 uptake or by an increase in biomass.

Rate of photosynthesis *increases with CO_2 concentration* but levels off at high concentrations.

Measuring carbon dioxide (CO_2) uptake is more complex and is usually done indirectly. For example, O_2 is released by aquatic plants during photosynthesis for (O_2) production.

Water pH rises when plants absorb carbon dioxide (CO_2) from water (HO_2), so measuring pH levels indicates CO_2 uptake.

 Copyright © Sterling Education

Abiotic factors of photosynthesis

Four *abiotic* (or *nonliving*) factors necessary for photosynthesis:

1) carbon dioxide (CO_2),

2) water (H_2O),

3) light and

4) temperature

These components influence the efficiency of photosynthesis.

Water is an *abiotic factor* that may or may not be plentiful at the location of an individual plant.

Water in photosynthesis is the source of O_2 *gas byproduct.*

Limiting factors of photosynthesis

Usually, only one is the *limiting factor* for a plant at a given time.

Earth's atmosphere contains 78% nitrogen and 21% oxygen, with the remaining 1% being a mixture of gases such as carbon dioxide.

Atmospheric CO_2 reaches photosynthetic tissues via *stomata* (regulated openings) on the underside of plant leaves.

CO_2 dissolves in a thin film of water covering the outside leaf cells and diffuses through cell walls to reach the chloroplasts.

Less than 1% of the water plants absorb is used for photosynthesis; the remainder is transpired (evaporated from leaves) or incorporated into cellular components.

In hot, dry climates, plants close stomata to conserve water, reducing CO_2 to the chloroplasts.

Photosynthesis exchanges CO_2 and O_2

Plant photosynthesis requires the intake of carbon dioxide (CO_2) and water (H_2O).

Water is absorbed by roots, moving water through vascular tissue in stems to leaves.

Mesophyll cells within the leaves are specialized, where photosynthesis takes place.

Exchange of O_2 (oxygen) and CO_2, and some water, occurs through the *stomata.*

Guard cells regulate stomata.

Stomata density depends on humidity and CO_2 concentration.

Any duplication (copies, uploads, PDFs) is illegal.

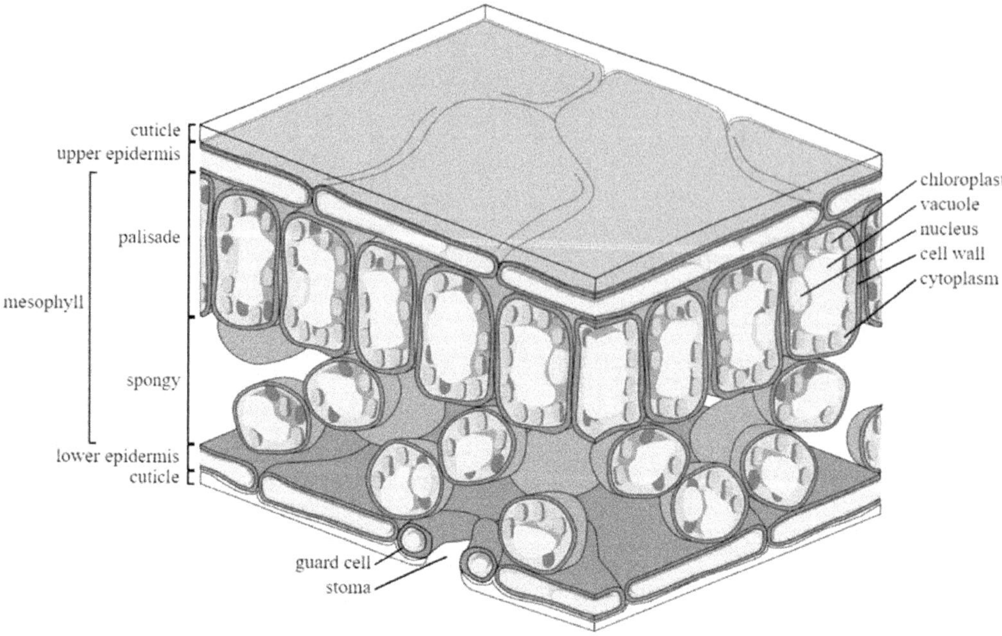

Cross-section of a typical C_4 plant leaf anatomy with guard cells regulating stomata aperture

Leaf modifications

Leaf undersides are often covered with hairs called *trichomes.*

Trichomes retain water, reduce airflow, and produce wax.

Waxy cuticle covers leaves outer epidermis, reducing water loss but limiting gas exchange.

Upper and lower leaf epidermis have *stomata* and serve a protective function.

Loosely arranged *spongy mesophyll* tissue creates air spaces.

Palisade mesophyll is tightly packed and contains the highest concentration of chloroplasts, which usually remain near the cell wall for optimal use of light.

Copyright © Sterling Education

Physical and Chemical Aspects of Photosynthesis

Solar energy

Solar radiation has a range of wavelengths, most filtered when passing through the atmosphere.

Ozone layer filters the high-energy wavelengths (e.g., gamma rays, x-rays, UV radiation).

Low-energy wavelengths (e.g., radio waves, microwaves, and infrared radiation) are filtered. 42% of the light from the sun hitting the Earth's atmosphere reaches the surface.

Visible light is most of the solar radiation reaching Earth's surface.

Photons

Photons are discrete particles (or *packets*) of light, named by Albert Einstein (1879-1955).

Photons are *both* a particle and a wave, exhibiting properties attributed to each.

Photon energy varies inversely with wavelength. Longer-wavelength light has lower energy, while shorter-wavelength light has higher energy.

Excitation is when molecules absorb light energy, and energy levels of electrons are elevated.

Light energy may be converted into many forms, a property photosynthesis exploit.

Fluorescence is when energy is emitted upon absorption.

Phosphorescence is when energy is emitted after a delay.

Visible light spectrum

Wavelength and energy are *inversely proportional.*

Therefore, red has the *longest wavelength* but the *lowest energy* in the visible spectrum.

Light is a crucial component of photosynthesis. Photosynthetic pigments utilize visible light, with photosynthesis rates varying by wavelength.

Light intensity depends on temperature, season, altitude, latitude, and atmospheric conditions.

Of the visible light reaching a leaf, approximately 80% is absorbed.

Violet-blue light (400 to 525 nm) and orange-red light (625 to 700 nm) are wavelengths most often absorbed for photosynthesis.

Light wavelengths within the visible light spectrum, ranging from red light at 780 nm to violet light at 390 nm, are solar radiation useful for photosynthesis.

For the visible light spectrum, mnemonic ROY G BIV (red, orange, yellow, green, blue, indigo, violet) describes the order of colors by decreasing wavelength (or increasing energy).

Any duplication (copies, uploads, PDFs) is illegal.

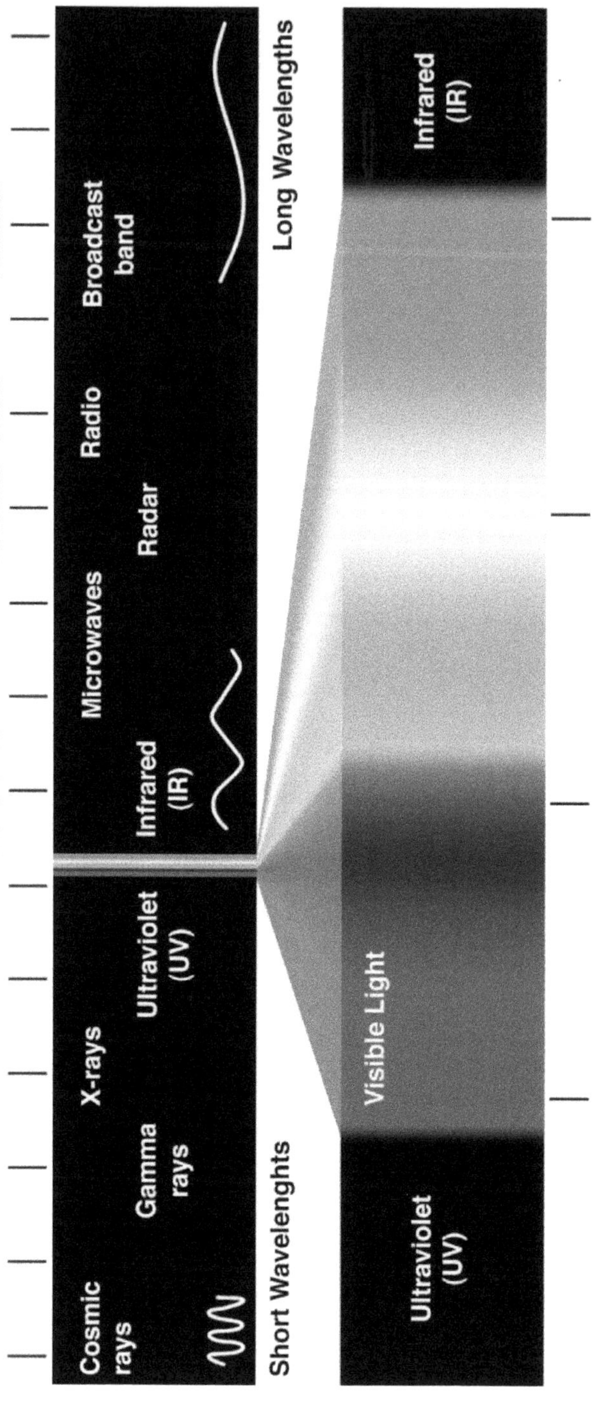

Electromagnetic spectrum includes visible light, with blue light having the most energy

Copyright © Sterling Education

Photolysis

Light intensity is a limiting factor for photosynthesis because, without sunlight, *photolysis* (splitting of water by photons) cannot occur during light-dependent reactions.

Photolysis produces a shortage of ATP and NADPH, products of the light-dependent reactions necessary for the Calvin cycle.

At low and medium light intensity, the rate of photosynthesis is *proportional to light intensity*.

High-intensity light is not necessarily beneficial, decreasing plant photosynthetic efficiency.

Temperature, like abiotic factors, has an optimum range. Enzymes for photosynthesis work slowly and less efficiently at low temperatures.

As temperature increases, photosynthesis rate increases steeply until the optimum temperature. If the temperature increases beyond this point, then the rate of photosynthesis decreases rapidly.

Any duplication (copies, uploads, PDFs) is illegal.

Notes for active learning

Copyright © Sterling Education

Photosynthetic Structures

Chloroplasts

CO_2 and H_2O enter a *mesophyll cell* and diffuse into the *chloroplast*, the site of the light-dependent reactions and the Calvin cycle.

Chloroplast is a double-membraned organelle known as a *plastid.* Each has ribosomes and identical copies of a double-stranded, circular DNA molecule unique to the cell's DNA.

Chloroplast membrane structures include an outer plasma membrane, an intermembrane space, and an inner plasma membrane.

Within the inner membrane, a fluid-filled space called *stroma* is the site of the Calvin cycle.

Thylakoids and light-dependent reactions

Thylakoids are flattened *disc-like sacs* organized into stacks as *grana* within the stroma.

Thylakoids have a *large surface area* for *light absorption.*

Inner lumen is where H^+ accumulates.

Protons (H^+) in the *thylakoid lumen* create an *electrochemical gradient* for photosynthesis.

Light-dependent reactions occur in the *thylakoids.*

Chlorophyll and pigments for solar energy absorption are *embedded in thylakoid membranes.*

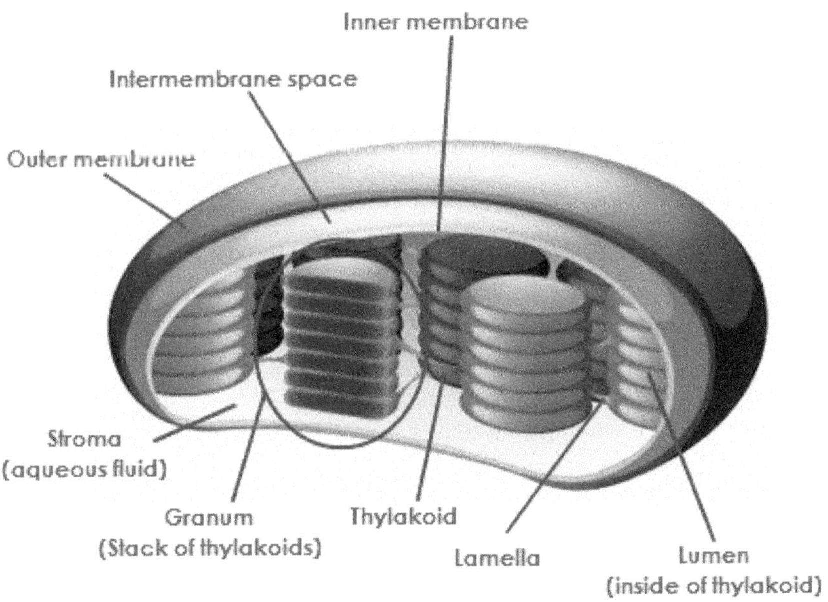

Chloroplast is a double-membered structure with stroma as the site of the Calvin cycle

Photosynthetic pigments

Pigment molecules have a distinctive *light absorption spectrum,* graphing the *percent light absorbed vs. wavelength.*

Action spectrum graphs plot photosynthetic *activity vs. wavelength of light.* Action spectrums (resembling absorption spectrums) indicate chlorophyll as primary pigment in photosynthesis.

Chlorophyll is the most abundant photosynthetic pigment in plants, so the rate of photosynthesis is highest at the low wavelengths of visible light (400 to 525 nm).

Chlorophyll absorbs light energy

Wavelengths comprising violet and blue light are the most readily absorbed by chlorophyll.

Chlorophyll absorbs an appreciable portion of red and orange light, which is why significant photosynthetic activity is found there.

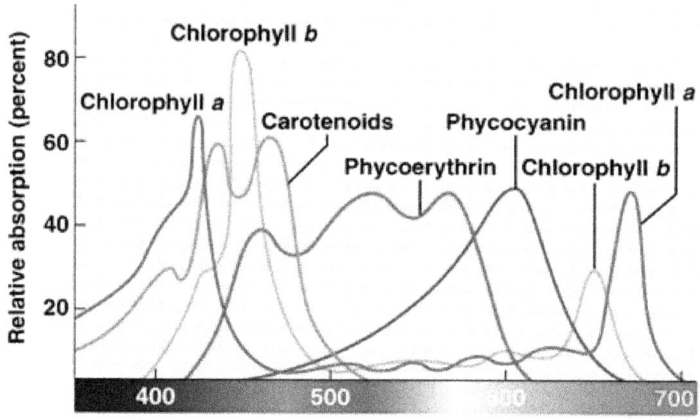

Relative absorption of visible light wavelengths (nm) by plant photosynthetic pigments

Little light is absorbed by chlorophyll at wavelengths of yellow and green light (525 to 625 nm), and most are reflected, giving plants their green color.

Photosynthetic organisms that are not green utilize alternate photosynthetic pigments. For example, red algae use *phycobilins,* which absorb blue, yellow, and green light but reflect red.

Chlorophyll molecules have a long lipid tail anchoring them in the lipid layers of thylakoid membranes, a *porphyrin ring* of alternating double and single bonds (conjugated system), and a single atom of magnesium in the center.

Chlorophyll is analogous to *heme,* the iron-containing pigment in hemoglobin protein within the red blood cells, which transports O_2.

 Copyright © Sterling Education

Chlorophyll *a* and chlorophyll *b*

Chloroplasts include chlorophyll, which performs *photosynthetic reactions.*

Chlorophyll exists in several forms, which differ slightly in molecular structure.

Chlorophyll a and *b* are common, with chlorophyll *a* three times as abundant as chlorophyll *b*.

Chlorophyll *a* ($C_{55}H_{72}MgN_4O_5$) is a bluish-green pigment.

Chlorophyll *b* ($C_{55}H_{70}MgN_4O_6$) is a yellow-green pigment.

Chlorophyll *b* broadens the spectrum of light available for photosynthesis.

Chlorophyll *b* absorbs light energy and transfers energy to chlorophyll *a* molecules.

Carotenoids and accessory pigments

Green plants have other pigments (carotenoids, flavonoids, betalains) for photosynthesis.

Carotenoids are yellow-orange pigments absorbing violet, blue, and green spectrums.

When chlorophyll decomposes in autumn, plant leaves turn yellow, orange, and red as carotenoids become visible.

Carotenoids are present in lesser amounts and allow a low rate of photosynthesis at wavelengths of light that chlorophyll cannot absorb.

Accessory pigments (antenna pigments) are light-absorbing pigments that aid chlorophyll *a*, including chlorophyll *b*, *c*, *d,* and carotenoids.

Photosystems and light-harvesting complexes

Light-harvesting complexes (or *antenna complexes*) contain 250 to 400 antenna pigment molecules and other compounds.

Light-harvesting complexes surround *reaction center,* a special pair of chlorophyll *a* molecules.

Photosystems have *antenna complexes* and *reaction centers* of thylakoid membranes.

Thylakoids are the structured membrane compounds that are present in chloroplasts.

Grana is a *stack of thylakoid membranes* in thylakoids.

There are many photosystems throughout the grana (granum, singular).

Photosystem II (PS II) and *Photosystem I* (PS I) in plant chloroplasts initiate photosynthesis.

Any duplication (copies, uploads, PDFs) is illegal.

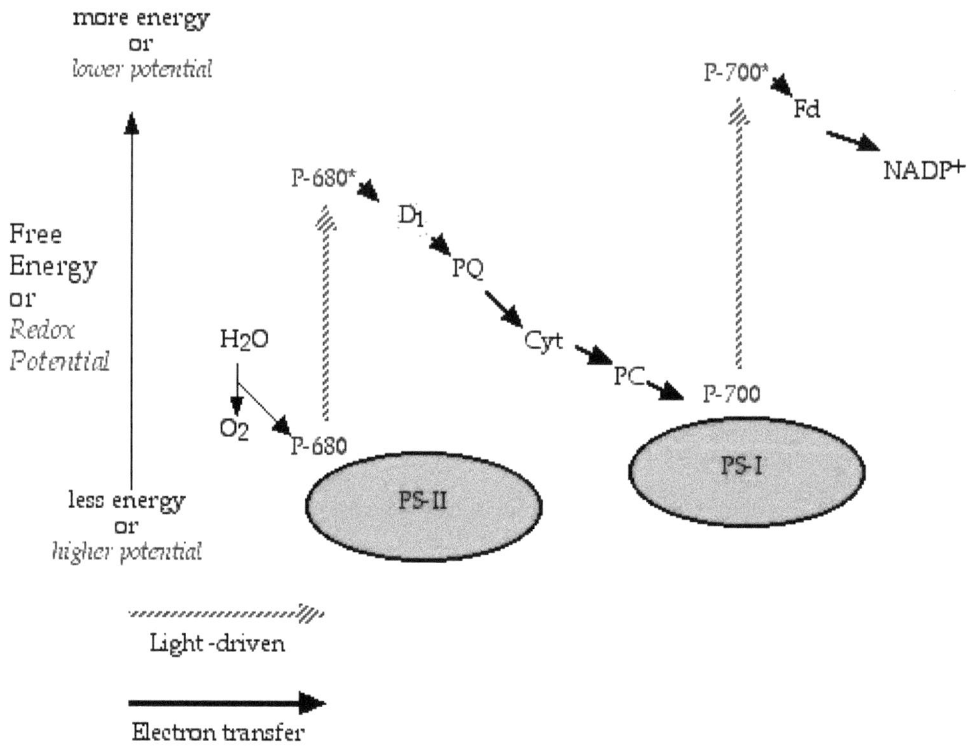

Photosystems II and I (light and electrons move in this order) are in the thylakoid membranes. Light-driven reactions include electron promotions for P-680 and P-700 with $NADP^+$ as an electron carrier.

 Copyright © Sterling Education

Photosynthetic Reactions

Light energy conversions

Light-dependent reactions require solar energy for energy-capturing reactions.

Light-dependent reactions *trap solar energy*, storing it as chemical energy (ATP, NADPH).

Light-dependent reactions begin with light striking *chlorophyll* molecules.

Subsequent reactions convert light energy to chemical energy (e.g., glucose, fructose).

Solar energy excites electrons in the antenna complex; it is *photoactivated.*

Energized electrons pass from one pigment molecule to the next until they reach the special chlorophyll *a* pair at the reaction center.

Electron transport chain

Electron transport chain (ETC) is reaction centers passing excited electrons to the first in a series of electron carriers (e.g., NADPH).

Excited electrons in the chlorophyll are unstable and re-emit absorbed energy as they migrate along the *electron transport chain.*

Electron transport chain produces ATP and NADPH.

Light reactions are from the initial photoactivation of a photosystem to ATP and NADPH synthesis. They are constant during daylight as plants make ATP and NADPH for Calvin cycle.

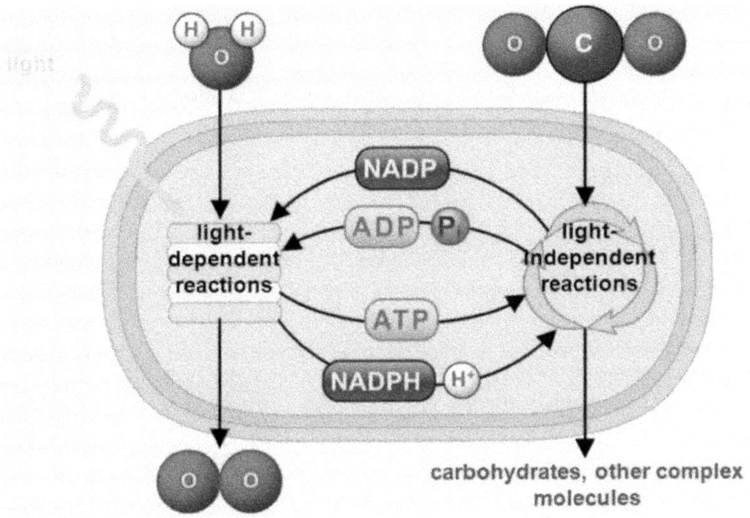

Light-dependent and light-independent (Calvin cycle) reactions during photosynthesis

Any duplication (copies, uploads, PDFs) is illegal.

Photosystems I and II

Two photosystems play critical roles in the light-dependent reactions of photosynthesis.

Photosystem I (PS I) is best excited by 700 nm light.

Photosystem II (PS II) cannot absorb photon energy of wavelengths longer than 680 nm.

Each photosystem has an antenna complex composed of chlorophyll, carotenoids, accessory proteins, and cofactors such as magnesium and calcium.

Oxidation-reduction reactions

Oxidation-reduction reactions occur throughout photosynthesis and cellular respiration.

Photosynthesis proceeds to the Calvin cycle reactions in the stroma, where NADPH and ATP created by the light-dependent reactions reduce CO_2 into carbohydrates.

Oxidation is the net loss of electrons, while reduction is the net gain of electrons.

Mnemonic "OIL RIG" for Oxidation Is Loss and Reduction Is Gain.

Light-dependent and light-independent reactions

Photosynthesis occurs in two stages.

> *Light-dependent reactions* capture light energy and use it to make energy storage and transport molecules ATP and NADPH.
>
> *Light-independent reactions* (Calvin cycle) use energy from short-lived electronically excited electron carriers (e.g., NADH, NADPH) to convert CO_2 and water into organic compounds.

Carbon fixation is this set of organic synthesis reactions.

Calvin cycle has four steps:

1) carbon fixation
2) reduction phase
3) carbohydrate formation
4) regeneration phase

ATP and NADPH provide energy for the chemical reactions in this sugar-generating process.

 Copyright © Sterling Education

Sunlight drives photosynthesis reactions

Sunlight provides the energy for the reaction:

ATP + NADPH + carbon dioxide (CO_2) → sugar

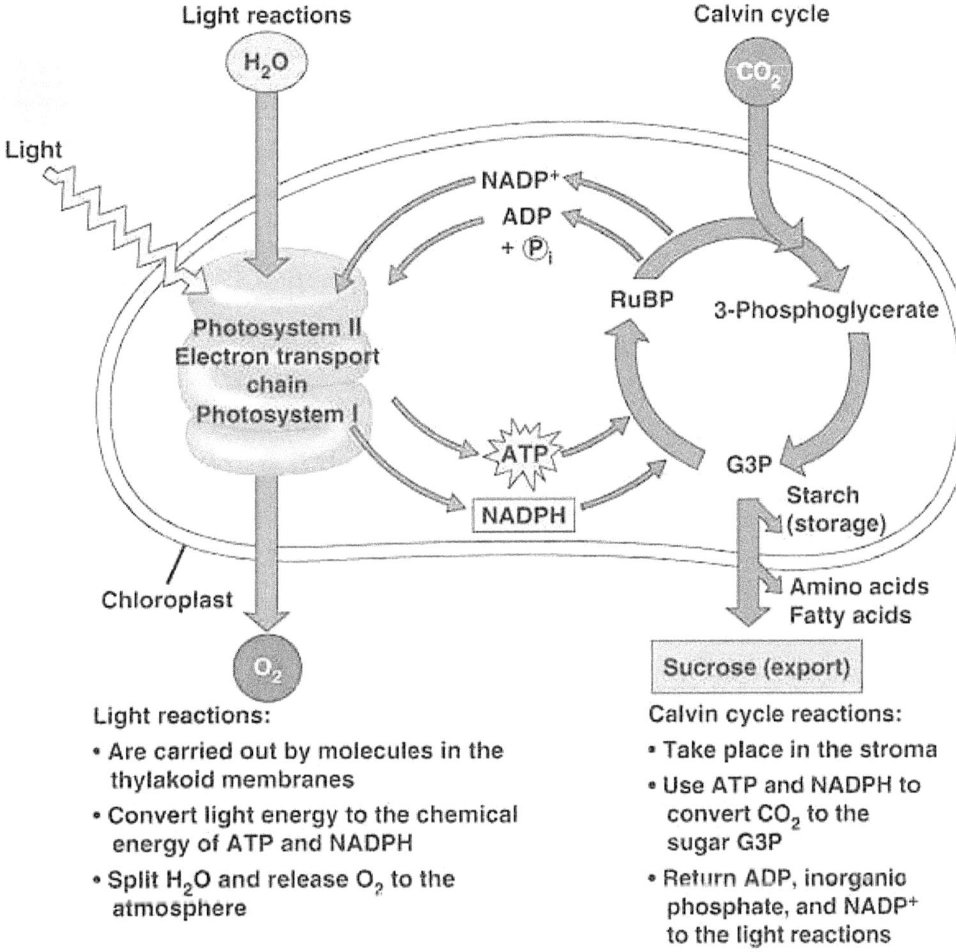

Products of light-dependent (ATP and NADPH) and light-independent reactions (sugars)

Any duplication (copies, uploads, PDFs) is illegal.

Photosynthesis uses oxidation–reduction reactions

Photosynthesis uses oxidation-reduction reactions:

- Photosynthesis is a redox reaction: water is oxidized to oxygen, and CO_2 is reduced to sugar and water.
- During photolysis, water is oxidized by *solar energy,* transferring from photons to oxygen, protons, and electrons.
- During the electron transport chain (ETC), each molecule is *reduced* (*gain electrons*) as it receives an electron and *oxidized* (*lose electrons*) as it passes the electron.
- Electron transport chain reduces $NADP^+$ to NADPH.
- Calvin cycle, CO_2 is reduced into sugar (carbohydrates) and water.

Light-dependent reactions

Light-dependent reactions in the *thylakoid membrane*:

noncyclic electron pathway and

cyclic electron pathway.

Each pathway produces ATP, but only the noncyclic electron pathway produces NADPH.

Photophosphorylation is ATP production via photosynthesis.

Cyclic and *noncyclic* pathways are *photophosphorylation.*

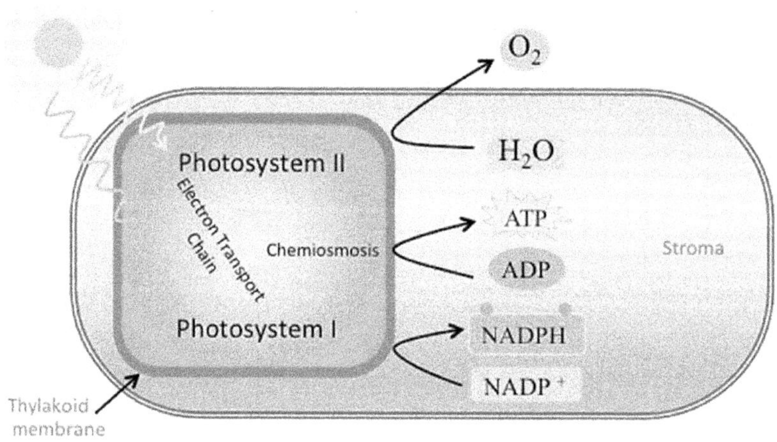

Light-dependent reactions include photosystems II and I that produce O_2, ATP, and NADPH

Copyright © Sterling Education

Noncyclic Electron Pathway

Noncyclic electron pathway

Noncyclic electron pathway is a light-dependent reaction in thylakoid membranes and requires the participation of PS I and PS II.

Noncyclic electron pathway begins at PS II, containing chlorophyll *a*, a carotenoid called *beta-carotene*, several electron-carrier proteins, and numerous cofactors.

PS II reaction center has a special pair of chlorophyll *a* molecules known as *P680.*

Noncyclic electron pathway begins when a photon strikes accessory pigments surrounding P680 near the inner surface of a thylakoid membrane.

Light energy excites the electrons of these pigments, transferring energy to the reaction center to excite electrons in P680.

Up to four photons strike PS II, but P680 only accepts one electron at a time, an unstable reaction with most energy lost to heat.

Pheophytin

Once excited, the electron (e^-) leaves the chlorophyll and passes from one molecule to the next along the electron transport chain.

Pheophytin, the *primary electron acceptor,* is the first molecule in the chain.

Pheophytin accepts the excited electron and transfers it to the *plastoquinone complex.*

Plastoquinone delivers e^- to *cytochrome complex*, passed to *plastocyanin*, and finally, PS I.

Oxidation-reduction reactions transfer electrons

Electrons (e^-) pass through a chain of oxidation-reduction reactions.

After each molecule in the electron transport chain passes an electron, it is reoxidized, readying it to accept a new electron from the molecule behind it in the chain.

However, it must be resupplied for P680 to send electrons down the chain continually.

Water splitting through photons (i.e., photolysis) releases electrons.

Any duplication (copies, uploads, PDFs) is illegal.

Oxygen-evolving complex

Photolysis occurs in an *oxygen-evolving complex* (OEC) with *manganese* and *enzyme complex*.

Photolysis produces 4 electrons, 4 protons, and O_2.

Four electrons are transferred to P680.

Protons remain in the thylakoid lumen.

O_2 is released into the atmosphere as a waste product of this process.

As electrons travel along the electron transport chain, protons (H^+ ions) are pumped across the thylakoid membrane into the thylakoid lumen, creating a *proton (or chemiosmotic) gradient.*

Protons travel across the membrane, *down concentration gradient*, but must pass through *ATP synthase* (enzyme complex embedded in the inner membrane), generating ATP (*oxidative phosphorylation*) with oxygen as the ultimate electron acceptor.

Noncyclic photophosphorylation

Synthesis of ATP via the noncyclic electron pathway is *noncyclic photophosphorylation.*

PS I reaction center is *P700*; a special pair of chlorophyll a molecules.

PS I activity resembles PS II, with chlorophyll a, beta-carotene, electron carriers, and cofactors.

PS II and PS I pigments absorb photons, but only reaction center molecules use light energy.

Pigments are accessory molecules to harvest light energy and pass it to the reaction center.

As in PS II, a photon of light strikes PS I, exciting electrons in P700.

Excited electrons pass along the electron transport chain.

Electrons continually resupply P700 shuttled along the electron transport chain before it.

 Copyright © Sterling Education

Ferredoxin

Primary electron acceptor after P700 is a series of iron-sulfur molecules *4Fe-4S.*

4Fe-4S passes electrons to the acceptor molecule, *ferredoxin* (NADP).

Ferredoxin releases electrons to *ferredoxin-NADP reductase* (FNR), catalyzing the reduction of $NADP^+$ to NADPH.

Reduction requires two electrons and one proton ($NADP^+ + 2\ e^- + H^+ \rightarrow NADPH$).

NADPH provides hydrogen (H^+) ions to the second series of major photosynthetic reactions: the Calvin cycle or *carbon-fixing reactions.*

Linear electron flow

Electron flow in the noncyclic pathway is *linear*:

> Photon strikes antenna complex of PS II → 4 electrons are released → P680 → Pheophytin → Plastoquinone → Cytochrome complex → Plastocyanin → P700 → 4Fe-4S → Ferredoxin → 2 NADPH

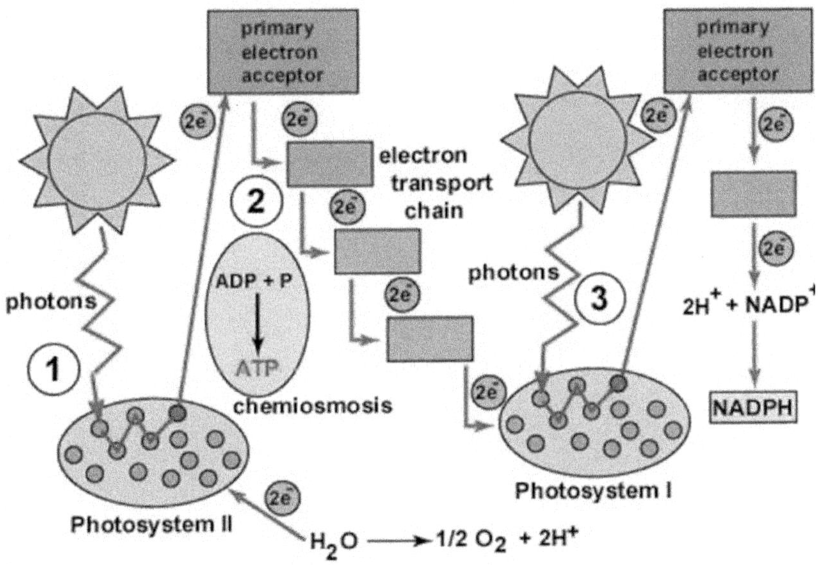

Light-dependent reaction with photosystem II and photosystem I as reactive centers to absorb electrons

Notes for active learning

Copyright © Sterling Education

Cyclic Electron Pathway

$NADP^+$ limits reactions

An organism may have the reductive power (NADPH) to synthesize carbon structures but needs ATP to power cellular activities in the chloroplast.

$NADP^+$ is needed for electron flow in thylakoid membranes, as it is the *final electron acceptor*.

If light intensity is not limiting, there usually is a shortage of $NADP^+$ as NADPH accumulates within the stroma.

If $NADP^+$ is unavailable, electron (e^-) flow is inhibited, promoting the cyclic electron pathway and bypassing NADPH production for ATP synthesis only.

Cyclic electron pathway when P700 receives an e^- from plastocyanin and passes it to ferredoxin.

However, because $NADP^+$ is unavailable, ferredoxin does not pass electrons to $NADP^+$.

Instead, it returns the electrons to the cytochrome complex, causing the *cytochrome complex* to pump protons (H^+) across the thylakoid membrane and create a proton (*proton-motive force*) gradient to drive ATP synthesis.

Electrons are returned to P700 via plastocyanin, repeating the process.

P700 cyclic pathway

PS I receives an electron not from PS II but from itself in the *cyclic electron pathway*.

Cyclic process as electrons return to PS I rather than move to $NADP^+$; PS I produces only ATP.

Electrons must be recycled continuously.

PS I as an ATP producer may seem counterintuitive since it is responsible for *NADPH* production in the *noncyclic electron pathway*, while PS II produces *ATP*.

However, PS I is excellent at transferring electrons.

PS I is a powerful reducing agent (electron donor), passing an electron to ferredoxin rather than $NADP^+$ to produce ATP when NADPH levels are high.

Any duplication (copies, uploads, PDFs) is illegal.

Chemiosmosis

Noncyclic and cyclic e^- pathways make ATP by *proton gradient* across thylakoid membrane.

Chemiosmosis is the movement of protons (H^+) *down a concentration gradient* across a selectively permeable membrane.

Chemiosmosis drives *ATP synthesis*.

Cyclic photophosphorylation is driven by the cyclic pathway between PS I and cytochrome.

Proton-motive force produced as electrons travel from P700 to ferredoxin to cytochrome causes cytochromes to pump protons into the lumen.

Proton gradient across inner membranes

Thylakoid lumen acts as a reservoir for H^+ ions.

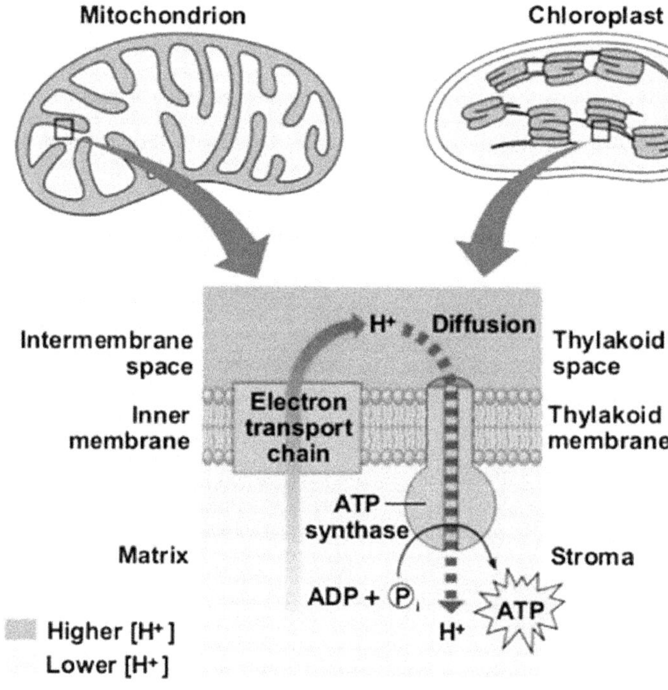

Chemiosmotic gradient comparison for cellular respiration (in mitochondria) and photosynthesis (in chloroplast)

Concentration gradient created by the accumulation of H^+ in the lumen causes H^+ to move back across the thylakoid membrane to the stroma, *down the concentration gradient*.

From chemiosmosis, ATP synthase uses the energy released from the movement of hydrogen ions down their concentration gradient to synthesize ATP from ADP and inorganic phosphate.

 Copyright © Sterling Education

Noncyclic photophosphorylation

Noncyclic photophosphorylation is a more complex process.

Hydrogen ions (*protons*) enter the thylakoid lumen at two points in *noncyclic electron pathway*.

> *First entry* in *noncyclic electron pathway* of protons into the thylakoid lumen occurs during photolysis: 2 H_2O split into O_2, 2 electrons, 2 H^+.
>
> $$2\ H_2O \rightarrow O_2 + 2\ \text{electrons} + 2\ H^+$$

Electrons are donated to P680, which resides in the transmembrane PS II complex.

Oxygen byproduct

H^+ is released into the thylakoid lumen; O_2 is a waste product released into the atmosphere.

Electrons travel from P680 to *pheophytin* to *plastoquinone* and reach the *cytochrome complex*.

> *Second entry* in the *noncyclic electron pathway* of protons into the thylakoid lumen occurs by electron transfer from plastoquinone to cytochrome via an intermediate carrier molecule, creating energy that pumps H^+ into the lumen from the stroma.

As cyclic photophosphorylation, H^+ moves *down* its gradient and *drives ATP synthase*.

Hydrogen (H^+) ions return to stroma for NADPH from $NADP^+$ and 2 electrons from PS I.

Any duplication (copies, uploads, PDFs) is illegal.

Notes for active learning

Copyright © Sterling Education

Calvin Cycle

Carbon-fixing reactions

Carbon-fixing reactions describe Calvin cycle; reactions have been light-independent (or dark).

ATP and NADPH are essential to light-dependent reactions and synthesize carbohydrates from atmospheric CO_2.

These reactions *do not* typically occur at nighttime. They indirectly require light because they rely on light-dependent reactions that require daylight. "Dark reaction" is less common, and "light-independent" is a misnomer.

Reactions occur in the stroma of chloroplasts if light-dependent reaction products are available.

Sugar synthesis

Depending on the plant, the carbon-fixing reactions may progress differently. Commonly, atmospheric CO_2 combines with the 5-carbon sugar *ribulose-1,5-bisphosphate* (RuBP).

CO_2 and RuBP convert into 6-carbon sugars, such as glucose, in several steps.

Some sugars combine into polysaccharides (e.g., starch) for storage within the plant.

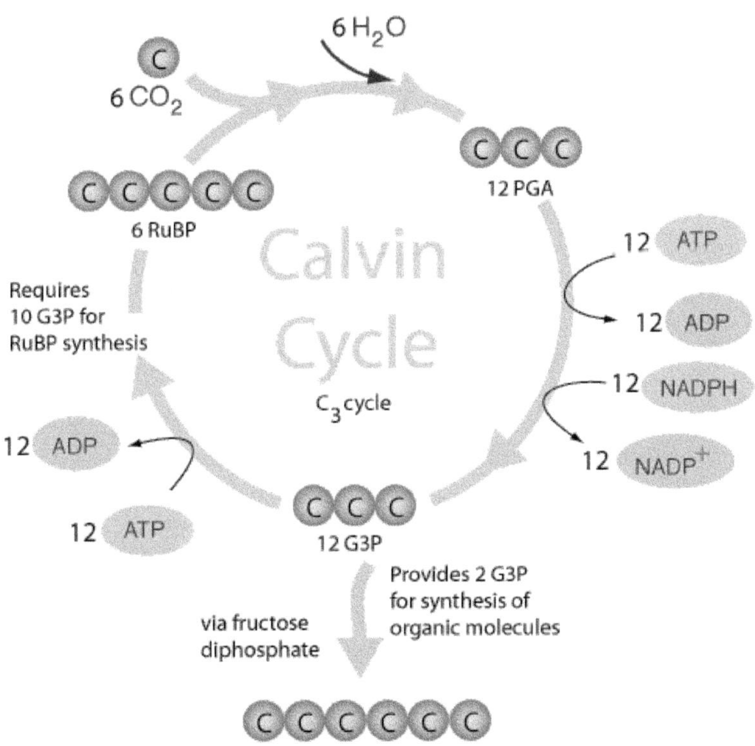

Calvin cycle reaction (light-independent) synthesizes sugars from CO_2.

Any duplication (copies, uploads, PDFs) is illegal.

Calvin cycle stages

Calvin cycle has three stages:

1) carboxylation

2) reduction

3) regeneration

Carboxylation reactions

1) Carboxylation (or fixation of carbon dioxide):

$CO_2 + RuBP \Rightarrow PGA$

Carboxylation attaches CO_2 to 5-carbon RuBP, forming an unstable 6-carbon intermediate.

Ribulose-1,5-bisphosphate carboxylase/oxygenase (RuBisCO) catalyzes reactions.

RuBisCO comprises 20–50% of the protein content of chloroplasts.

RuBisCO is the most common protein in nature and is required for most carbon fixing (*synthesis of sugars*) reactions.

When 6-carbon intermediate forms, it splits to form two molecules of the 3-carbon compound *3-Phosphoglyceric acid* (3-PG, 3-PGA, or PGA).

3-Phosphoglyceric acid is *glycerate 3-phosphate*.

Reduction reactions

2) Reduction of 3-phosphoglyceric acid:

$PGA + ATP + NADPH \Rightarrow G3P + ADP + Pi + NADP+$

PGA is phosphorylated into an intermediate by ATP produced in light-dependent reactions.

Intermediate is reduced by NADPH (other product of light-dependent rxn) to *glyceraldehyde 3-phosphate*, a 3-carbon compound like PGA.

Glyceraldehyde 3-phosphate (G3P, GP, GA3P, or GAP) is *triose phosphate* (TP) and *3-phosphoglyceraldehyde* (PGAL).

G3P reduction makes ADP, inorganic phosphate (Pi), and $NADP^+$.

G3P is converted into useful biomolecules, such as glucose. Glucose synthesizes polysaccharides (i.e., starch and cellulose). It combines with fructose, forming sucrose, a carbohydrate.

Copyright © Sterling Education

Regeneration reactions

3) Regeneration of RuBP:

G3P + ATP ⇒ RuBP + ADP + Pi

However, only one G3P can be used for conversion into glucose.

G3P (3-carbon) regenerates RuBP, which is essential for carbon fixation.

G3P engages a carbon acceptor and undergoes a series of reactions, requiring energy from ATP, converting them into RuBP.

At this point, the cycle begins again.

Description followed the fate of one original RuBP molecule; the process uses six RuBP and six CO_2 for each cycle.

Carboxylation converts six 5-carbon RuBP and six CO_2 into six 6-carbon unstable intermediates.

Six intermediates break down into twelve 3-carbon PGA.

6 RuBP + 6 CO_2 → 12 PGA

Calvin cycle summary reactions

Reduction:

12 PGA + 12 ATP + 12 NADPH → 12 3-carbon G3P + 12 ADP + 12 Pi + 12 $NADP^+$

Two G3P create *glucose phosphate.*

12 PGA + 12 ATP + 12 NADPH → 12 G3P + 12 ADP + 12 Pi + 12 $NADP^+$

During regeneration:

10 G3P, using 6 ATP, are remade into 6 RuBP, producing 6 ADP and 4 Pi.

Regeneration reaction:

10 G3P + 6 ATP → 6 RuBP + 6 ADP + 4 Pi

Any duplication (copies, uploads, PDFs) is illegal.

Notes for active learning

Copyright © Sterling Education

Types of Photosynthesis

C_3, C_4 and CAM photosynthesis

C_3, C_4, and CAM describe how plants perform photosynthesis.

C_3 and C_4 refer to the carbon length of the first photosynthetic carbohydrate product.

C_3 plants produce a 3-carbon molecule, 3-phosphoglyceric acid (PGA).

C_4 plants produce a 4-carbon molecule, oxaloacetate (OAA).

Some plants do not undergo the Calvin cycle. C_4 and CAM (*crassulacean acid metabolism*) plants produce oxaloacetate but undergo photosynthesis differently.

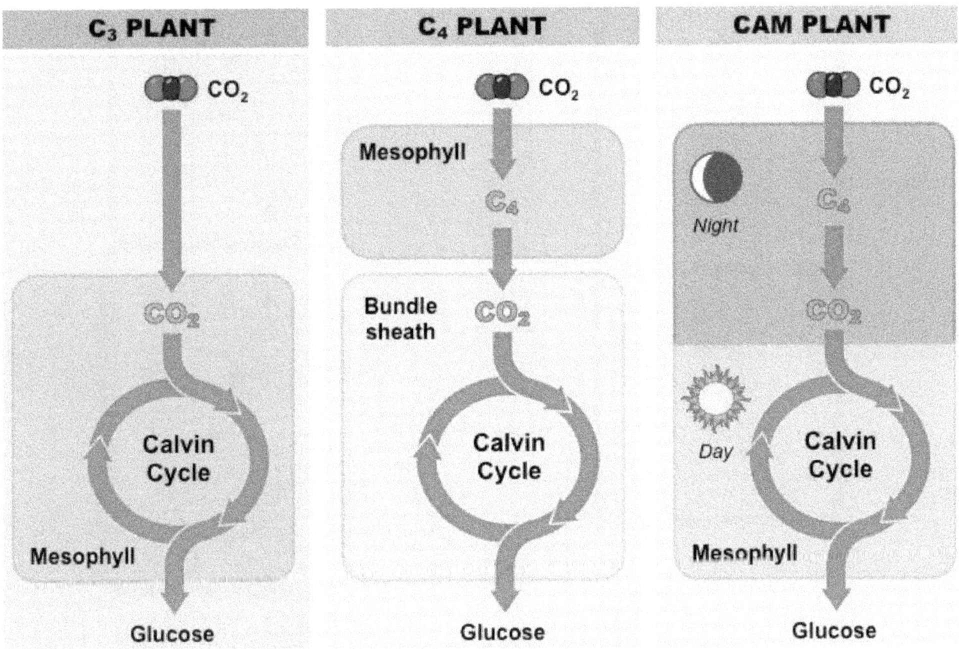

C_3, C_4 and CAM plants comparing glucose synthesis with mesophyll and bundle sheath cells

C_3 photosynthesis

More than 90% of angiosperms are C_3 plants.

In C_3 plants, the Calvin cycle proceeds, in which CO_2 is fixed into the 3-carbon molecule PGA.

Stomata close in heat to conserve water reducing CO_2 in leaves and increasing O_2.

High O_2 concentration is a disadvantage; RuBisCO uses O_2 in Calvin cycle rather than CO_2.

O_2, in this way, is *photorespiration* by intaking O_2 and producing CO_2, like cellular respiration.

Photorespiration

RuBP reacts with O_2 to create CO_2 during photorespiration.

In contrast to photosynthesis, when RuBP reacts with CO_2 to form carbohydrates (e.g., glucose).

Photorespiration is an *oxygenation* reaction.

Photosynthesis is a *carboxylation* reaction.

RuBisCO performs oxygenation rather than carboxylation around 25% of the time, which may be promoted by low CO_2, high O_2, or elevated temperatures.

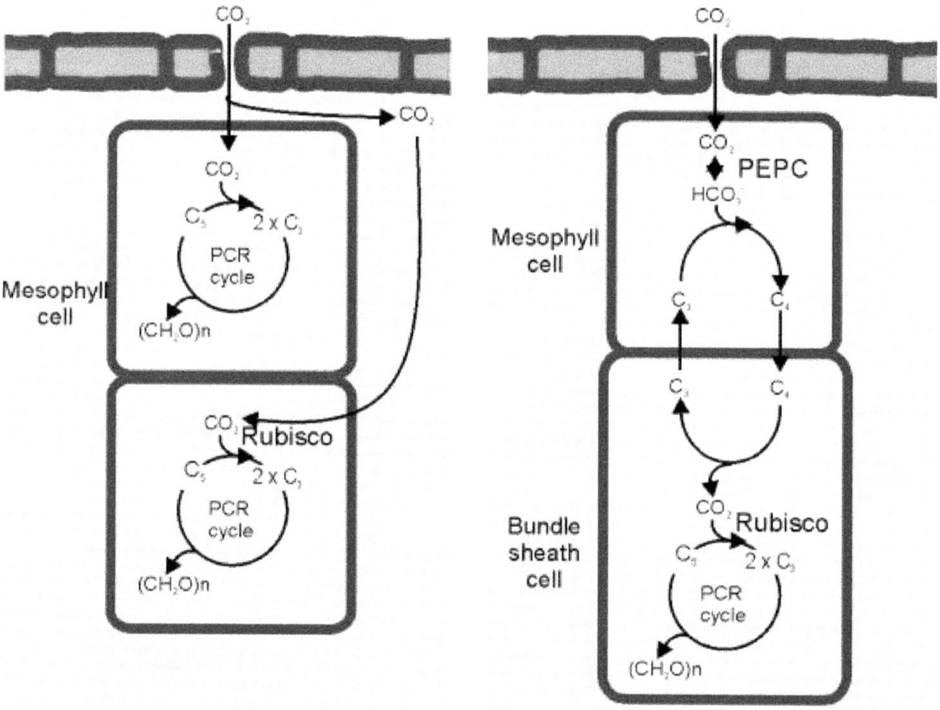

C_3 (left) and C_4 plants comparison for the fiixation of CO_2

Photorespiration produces one PGA and one *phosphoglycolate*.

Phosphoglycolate inhibits carboxylation.

Plants expend energy by converting phosphoglycolate and reclaiming two carbons.

Peroxisomes break down the products of this process.

 Copyright © Sterling Education

Oxygenation reaction

Photorespiration is wasteful and inefficient, resulting in a net energy loss for the plant. Plants expend up to 40% of the energy stored in sugars (e.g., glucose, fructose) for inevitable damage created by oxygenation reactions.

Photosynthesis evolved billions of years ago when the concentration of atmospheric O_2 was low and photorespiration was rare.

Photorespiration grew to be a great inconvenience to plants.

Oxygenation reaction developed as an adaptation to Earth's early atmosphere.

Plants evolved mechanisms to *reduce damage caused by* O_2 in the Calvin cycle.

Humans have attempted to circumvent issues created by the oxygenation reaction.

Carboxylation without oxygenation

Molecular genetics modified RuBisCO, eliminating oxygenation while retaining carboxylation. The two reactions have not been separated because modifications in RuBisCO reduce oxygenase and carboxylase activity.

Nature developed a system to avoid photorespiration.

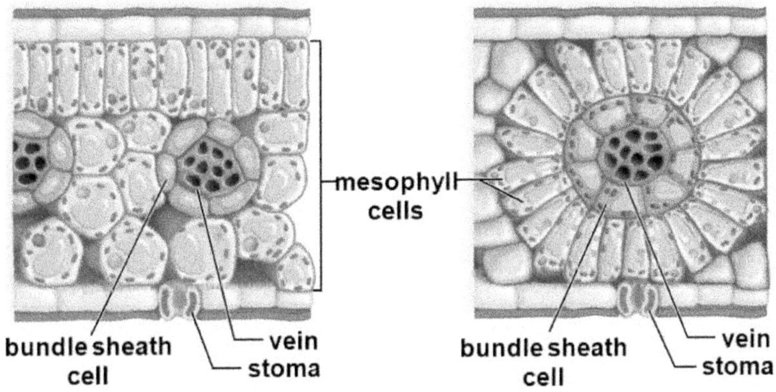

Anatomy of C_3 (left) and C_4 plants. C_4 photosynthesis uses carbon concentration to improve the efficiency of photosynthetic carbon fixation. Leaves of most C_4 plants have a Kranz leaf anatomy with a bundle sheath and mesophyll cells.

Any duplication (copies, uploads, PDFs) is illegal.

C_4 photosynthesis

Corn and sugar cane are important C_4 plants.

C_4 photosynthesis is in less than 5% of plants, mostly in hot, dry climates.

C_4 photosynthesis is exceptionally well represented in the grasses.

C_4 plants avoid photorespiration; the net photosynthetic rate is 2 to 3 times more than C_3 plants.

C_4 photosynthesis evolved from C_3 plants independently many times during evolution.

The enzymes co-opted for C_4 photosynthesis were present in C_3 plants from which they evolved.

However, in moist or cold environments, C_3 plants are more efficient.

Researchers attempt to artificially convert certain C_3 plants into C_4 plants to maximize crop yields and allow agriculture in hostile environments.

C_3 plants have chloroplasts in mesophyll cells, but C_4 plants have additional chloroplasts in special *bundle sheath cells*.

Bundle sheath cells surround the leaf veins, known as *Kranz anatomy.*

Mesophyll cells

In C_4 plant leaves, mesophyll cells are arranged concentrically around the bundle sheath cells.

Light-independent reactions are split between cell types, with:

Preliminary carbon fixation in *mesophyll cells* and

Calvin cycle in *bundle sheath cells*.

RuBisCO is in *bundle sheath cells*.

Rather than RuBisCO, mesophyll cells have an alternate phosphoenolpyruvate (PEP) enzyme.

CO_2 enters C_4 leaves and is absorbed by mesophyll cells, as in C_3 plants.

However, rather than being fixed by RuBisCO into PGA, CO_2 combines with PEP to form 4-carbon *oxaloacetate* (OAA).

PEP carboxylase (PEP-C or PEPCase) catalyzes this reaction.

PEPCase assimilates carbon more efficiently while avoiding photorespiration.

PEPCase has no affinity for O_2, unlike RuBisCO, and has a higher affinity for CO_2 at high temperatures than RuBisCO.

Copyright © Sterling Education

Bundle sheath cells

Oxaloacetate is reduced to *malate* and pumped into the bundle sheath cells.

Some plants reduce oxaloacetate to aspartate rather than malate.

Other plants do not alter oxaloacetate before shuttling it to the bundle sheath cells.

In bundle sheath cells, *malate* is converted to *oxaloacetate* and *decarboxylated* (CO_2 is removed) to form 3-carbon *pyruvate.*

CO_2 remains in the bundle sheath cells to begin the Calvin cycle.

Pyruvate is returned to the mesophyll cells to regenerate into PEP using ATP, and its efficiency is lost in cooler temperatures.

The initial fixed carbon form, whether oxaloacetate, malate, or aspartate, does not substitute for any of the carbon compounds of the Calvin cycle, such as RuBP or PGA.

These acids serve as sources of CO_2 for the conventional Calvin cycle in bundle sheath cells.

CO_2 is fixed into RuBP as normal and proceeds throughout the cycle.

In this regard, C_4 plants fix carbon twice: once in the mesophyll and once in the bundle sheath.

CAM photosynthesis

CAM (*crassulacean acid metabolism*) plants include cacti, stonecrops (family Crassulaceae, the strategy is named), orchids, bromeliads, and succulents.

CAM photosynthesis is an alternative to C_3, almost exclusively in plants from dry environments.

CAM (like C_4) plants use PEPCase to fix CO_2 into *oxaloacetate*, separate from the Calvin cycle.

While C_4 plants separate them spatially, with the former in *mesophyll cells* and the latter in *bundle sheath cells*, CAM plants separate them temporally.

Occurs in *mesophyll cells*, but initial *CO_2 occurs at night*; the *Calvin cycle during the day.*

Any duplication (copies, uploads, PDFs) is illegal.

Oxaloacetate and mesophyll

CAM plants have *stomata* opened only at night when CO_2 is taken in and placed in *mesophyll.*

CAM plants use PEPCase to react with CO_2 and PEP to form *oxaloacetate.*

Oxaloacetate converts to *malate* and stores as malic acid in large vacuoles in mesophyll cells.

During the day, *malic acid* returns to the chloroplasts for conversion into *oxaloacetate*

Oxaloacetate is decarboxylated (removing CO_2) into pyruvate and CO_2.

CO_2 is introduced into the Calvin cycle, while pyruvate is used to regenerate PEP.

Stomata conserve water

CAM evolved to avoid transpiration rather than reduce the effects of photorespiration, although it is a benefit.

CAM advantage over a C4 strategy is *conserving water* by *closing stomata* during the day.

No CO_2 intake during the daytime, as *stomata are closed* to *avoid transpiration.*

Photosynthesis in a CAM plant is minimal due to limited CO_2 fixed at night, allowing them to live in stressful conditions.

Copyright © Sterling Education

Notes for active learning

Notes for active learning

Copyright © Sterling Education

STERLING
Education

PRACTICE QUESTIONS
&
DETAILED EXPLANATIONS

Page intentionally left blank

Copyright © Sterling Education

Practice Questions: Plants

1. In angiosperms, pollen grains are produced in the:

A. ovule
B. sepal
C. stigma
D. carpel
E. anther

2. In most plants, which organs are adapted to capture sunlight for photosynthesis?

A. leaves
B. flowers
C. roots
D. stems
E. stomata

3. Which is true?

A. Plants take carbon dioxide from the atmosphere and do not use oxygen
B. Most plants get the water they need from the soil
C. Plants have adaptations that minimize light absorption
D. Plants require less water on sunny days
E. The leaves provide most of the water used by plants

4. In angiosperms, xylem consists of tracheids and:

A. vessel elements
B. parenchyma
C. sieve tube elements
D. albuminous cells
E. companion cells

5. In angiosperms, flowers are adaptations for reproduction. Flowers that are pollinated by animals do NOT:

A. produce nectar to attract pollinators
B. pollinate more efficiently than wind-pollinated plants
C. grow larger
D. exist as brightly colored structures
E. exist as small structures

Any duplication (copies, uploads, PDFs) is illegal.

6. Without gas exchange, a plant would be unable to:

A. make minerals
B. reproduce
C. make food
D. absorb sunlight
E. flower

7. Which is NOT part of the female structure of a flower?

A. ovary
B. stigma
C. style
D. filament
E. ovules

8. Living on land requires those plants:

A. have cell walls
B. exchange gases
C. conserve water
D. have photosynthetic pigments
E. utilize oxygen

9. If some xylem of a young maple tree were destroyed, it would interfere with the tree's ability to:

A. absorb nutrients from the soil
B. conduct water to the leaves
C. conduct water to the roots
D. conduct sugars to the roots
E. undergo photosynthesis

10. Vascular tissue in plants consists of:

A. epidermis
B. parenchyma and collenchyma
C. meristems
D. xylem and phloem
E. rhizoid

11. The ancestors of land plants likely evolved from:

A. prokaryotes that carried on photosynthesis
B. a protist that lived on land
C. an organism similar to green algae
D. mosses that lived in water
E. mosses that lived on land

Copyright © Sterling Education

12. When a plant reproduces vegetatively:

A. only root tissue can be used to produce new offspring
B. offspring are produced by mitosis alone
C. meiosis produces a new gametophyte
D. offspring will differ from the parent
E. mitosis produces a new gametophyte

13. Seeds that animals spread usually are contained in:

A. lightweight structures
B. thin coatings that are easily digested
C. unripen ovaries
D. environmentally resistant structures
E. fleshy and nutritious fruits

14. To observe mitosis, which should be examined under a compound microscope?

A. xylem from a tree trunk
B. phloem from the stem of a plant
C. epidermis of a leaf
D. tip of a shoot
E. rhizoid

15. What would have likely occurred if plants had not begun to live on land?

A. Animals would not undergo cellular respiration
B. There would be no CO_2 in the atmosphere
C. Animals also would not live on land
D. There would be no green algae in the oceans
E. There would be no oxygen in the atmosphere

16. The vascular cylinder of a root consists of the following:

A. phloem and xylem
B. phloem and ground tissue
C. xylem only
D. phloem only
E. xylem and ground tissue

17. Vegetative reproduction can be a disadvantage when:

A. the parent plant must reproduce quickly
B. the parent plant is not competing with other organisms for resources
C. conditions are favorable for growth
D. conditions in the physical environment suddenly change
E. the parent plant reproduces slowly

Any duplication (copies, uploads, PDFs) is illegal.

18. The layer of cells that encloses the vascular tissue in the central region of a root is:

A. apical meristem
B. epidermis
C. cortex
D. meristems
E. endodermis

19. What is the fundamental difference between a sporophyte and a gametophyte?

A. A sporophyte is in the diploid phase, while a gametophyte is in the haploid phase of the plant life cycle
B. A sporophyte is much smaller than the gametophyte
C. A sporophyte is a reproductive structure, while a gametophyte is not
D. A sporophyte undergoes sexual reproduction, while a gametophyte undergoes asexual reproduction
E. A gametophyte is in the diploid phase, while a sporophyte is in the haploid phase of the plant life cycle

20. The outer covering of a plant consists of:

A. meristematic tissue
B. dermal tissue
C. vascular tissue
D. ground tissue
E. albuminous cells

21. How have the sporophyte and gametophyte changed as plants have evolved?

A. The gametophyte has become smaller, while the sporophyte has become larger
B. Many more recently evolved plants no longer have a sporophyte
C. The sporophyte has become larger, while the gametophyte has not changed
D. More recently evolved plants are mostly haploid gametophytes
E. The gametophyte has become larger, while the sporophyte has become smaller

22. A period during which the embryo of a seed is alive but not growing is:

A. germination
B. dormancy
C. dispersal
D. fertilization
E. necrosis

Copyright © Sterling Education

23. Which correctly describes seeds and fruits in angiosperms?

A. Fruits provide nutrients for the seed
B. Fruits do not protect the seed, so they are not favored by natural selection
C. Fruits surround no more than one seed apiece
D. Fruits develop from the endosperm of the seed
E. Fruits are matured ovaries that contain seeds

24. All are reasons why green algae are classified as plants, EXCEPT:

A. They have genes like genes in land plants
B. They have chlorophyll *a*
C. They have cellulose in their cell walls
D. They are single-celled organisms
E. They have chlorophyll *b*

25. In bryophytes, which structure is like a root?

A. stalk
B. capsule
C. rhizoid
D. sporophyte
E. stomata

26. A seed that is dispersed to an area far away from the parent plant always faces less:

A. unfavorable conditions for germination
B. favorable conditions for germination
C. chance of self-pollination
D. competition for space
E. competition with the parent plants

27. Root pressure:

A. is produced in the vascular cylinder by osmotic pressure
B. is produced within the medulla of the root
C. is produced within the cortex of the stem
D. causes plant roots to increase in size
E. forces water in the xylem downward

Any duplication (copies, uploads, PDFs) is illegal.

28. Which statement about green algae is true?

A. Like other plants, they have specialized structures
B. They are multicellular plants
C. Evidence suggests they were the first plants
D. They are found in dry areas on land
E. none of the above

29. The plant embryo in a seed begins to grow again during:

A. pollination
B. germination
C. dormancy
D. fertilization
E. B and C

30. A period of dormancy can allow seeds to germinate:

A. in extreme temperatures
B. without water
C. under cold conditions
D. under ideal conditions
E. under warm conditions

31. One of the three main functions of stems is to:

A. store sugars
B. store carbon dioxide
C. facilitate photosynthesis
D. carry nutrients and H_2O between roots and leaves
E. store water

32. Which hormone stimulates an increase in size, especially in stems and fruits?

A. germination
B. ethylene
C. auxin
D. cytokinin
E. gibberellin

33. Of the following locations, mosses likely grow in:

A. exposed ground with direct sun
B. shaded ground near a small pond
C. sandy soil near a beach
D. hot and dry soil
E. mineral-poor soil

 Copyright © Sterling Education

34. When environmental conditions are unfavorable, why is it advantageous for green algae to reproduce sexually?

A. Sexual reproduction produces offspring with greater genetic variety and durable zygotes that help green algae survive unfavorable conditions
B. Sexual reproduction produces multicellular algae instead of unicellular algae. Multicellular algae are better able to survive harsh conditions
C. In unfavorable conditions, they are multicellular, which favors sexual reproduction
D. Sexual reproduction produces larger numbers of offspring in less time
E. None of the above

35. Many cacti have large stems and no leaves. Which leaves' function is assumed by the stems of these cacti?

A. absorb H_2O from the soil
B. transport materials throughout the plant
C. absorb nutrients from the soil
D. produce food by photosynthesis
E. store excess H_2O

36. Which vascular tissue carries solutions of nutrients and carbohydrates produced by photosynthesis?

A. lignin
B. tracheids
C. phloem
D. xylem
E. rhizoid

37. Unlike roots, stems:

A. may be involved in photosynthesis
B. are protected by epidermal cells
C. have ground tissue
D. transport water
E. transport sugars

38. During primary growth, a stem:

A. increases in width
B. fills with O_2
C. produces wood
D. produces flowers
E. increases in length

Any duplication (copies, uploads, PDFs) is illegal.

39. Hormones that stimulate cell elongation and are produced in rapidly growing regions near the tip of plant roots or stems are:

A. cytokinins
B. gibberellins
C. auxins
D. ethylene
E. sepals

40. What might a thin tree ring indicate?

A. decreased production of phloem
B. a year of drought
C. increased production of xylem
D. xylem production in winter
E. decreased production of xylem

41. What is the phenomenon affecting the growth of the bean shoot in the figure below?

A. thigmotropism
B. rapid movement
C. gravitropism
D. phototropism
E. gibberellin

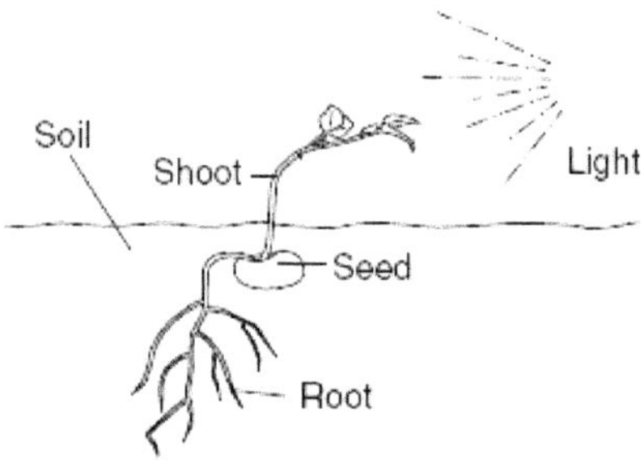

42. Bryophytes must live in moist areas without vascular tissue, and what other reason requires moist areas?

A. Carbon dioxide exchange is more efficient in wet areas
B. Without moisture, rhizoids cannot anchor the plants
C. Bryophytes need extra water for photosynthesis
D. Oxygen exchange is more efficient in wet areas
E. The sperm of bryophytes need water to swim to an egg

43. Which structure includes a plant embryo, a food supply, and a protective covering?

A. Seed
B. Gametophyte
C. Pollen grain
D. Spore
E. Xylem

Copyright © Sterling Education

44. When positioned horizontally, a plant's response to gravity is that:

A. all parts of the plant bend upward, away from the force of gravity
B. auxins in the shoot cause the growth of tendrils to pull the shoot upwards
C. auxins in the lower sides of stems cause cell elongation that bends the stem upright
D. the shoot bends to grow downward, toward the force of gravity
E. none of the above

45. What is the source of ethylene gas in a plant?

A. stems
B. fruit tissues
C. roots
D. leaves
E. all the above

46. The stomata of leaves are usually open:

A. at night if a plant has enough water
B. during the day if a plant has enough CO_2
C. during the day if a plant has enough water
D. during the day if a plant has too little water
E. at night if a plant has too little water

47. The yellow, orange, and red colors of leaves in the fall are the result of the following:

A. decrease of auxins in the meristems
B. production of anthocyanin and the breakdown of chlorophyll
C. new synthesis of orange and yellow pigments in leaves
D. movement of chlorophyll from the stems to the leaves
E. increase of phytochrome in the leaves

48. O_2 and CO_2 move in and out of a leaf through:

A. stomata
B. phloem
C. guard cells
D. palisade mesophyll
E. xylem

49. The growth of plant seedlings is usually:

A. gravitropic and influenced by ethylene
B. phototropic and influenced by ethylene
C. thigmotropic and gravitropic
D. phototropic and gravitropic
E. phototropic only

Any duplication (copies, uploads, PDFs) is illegal.

50. What is one reason vascular plants are larger than nonvascular plants?

A. Nonvascular plants do not reproduce as quickly as vascular plants do, so they do not grow as large
B. Vascular tissue carries water and nutrients much more efficiently than can be carried by diffusion
C. Vascular plants live in areas where they can get more water than nonvascular plants
D. Nonvascular plants cannot live in as many areas as vascular plants, so they stay small
E. The rate of gas exchange is lower in nonvascular plants, which accounts for their diminished size

51. Most of the photosynthesis in plants occurs in:

A. xylem
B. stomata
C. mesophyll tissue
D. guard cells
E. phloem

52. Stomata are only on the lower surface of leaves in many plants. The explanation for this is:

A. water loss would be less on the shaded lower surface than in the direct sun
B. gravity functions in gas exchange
C. stomata are closer to vascular bundles that bring water into the leaf
D. photosynthesis only occurs in the spongy mesophyll near the bottom of the leaf
E. photosynthesis only occurs in the regions surrounding the stomata

53. The growth of ivy tendrils that wrap around objects is an example of:

A. photoperiodism
B. chemotropism
C. phototropism
D. gravitropism
E. thigmotropism

54. Seed-bearing plants differ from all other plants in that:

A. they have true roots, stems, and leaves
B. their gametes do not require water for fertilization
C. they have a gametophyte generation
D. they have only xylem and no phloem
E. they have neither xylem nor phloem

Copyright © Sterling Education

55. Xylem tissue is vital because it:

A. carries out photosynthesis
B. carries carbohydrates to all parts of the plant
C. allows water to diffuse from the roots
D. can conduct water over long distances
E. carries minerals to the roots

56. Corn, broccoli, and cabbage were developed by:

A. auxins
B. selective breeding
C. plant propagation
D. germination
E. pollination

57. In the figure, the water pressure is:

A. high in the guard cells
B. low in the guard cells
C. low in the stoma
D. high in the stoma
E. more than one of the above

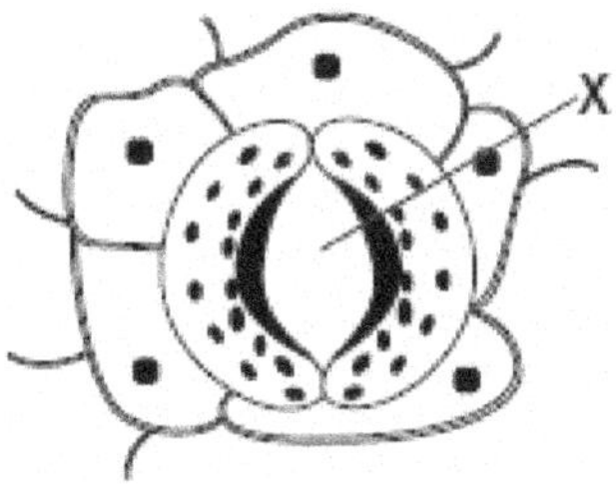

Underside of a leaf

58. Through which plant cell does water move by capillary action?

A. mesophyll cells
B. albuminous cells
C. phloem cells
D. guard cells
E. xylem cells

59. Which gymnosperm reproductive structure is haploid?

A. seed
B. cone scale
C. pollen grains
D. seed cone
E. embryo

60. Which BEST describes how people use plants?

A. All medicines come from plants
B. Plants break down decomposing organisms and waste
C. People use only plants for food
D. Plants provide food, raw materials, and several medicines
E. Plants release carbon dioxide into the environment

Any duplication (copies, uploads, PDFs) is illegal.

61. In the figure, which letter indicates the structure where the embryo develops?

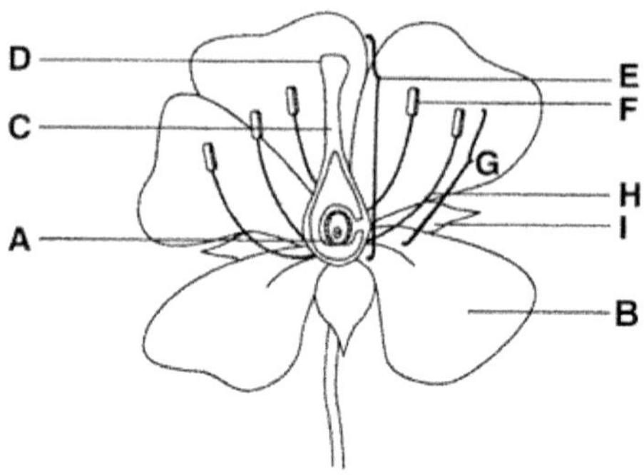

A. A
B. B
C. C
D. D
E. F

62. The movement of water containing food molecules in the phloem can be explained by

A. root pressure
B. the pressure-flow hypothesis
C. transpiration pull
D. diffusion gradient
E. capillary action

63. According to the pressure-flow hypothesis, which statement is NOT true?

A. The movement of H_2O into a nutrient-rich region of the phloem decreases the pressure in that region
B. Phloem can move sugars in either direction to meet the nutritional needs of the plant
C. H_2O is necessary for sugars to move through the phloem
D. H_2O moves from the xylem to the phloem of a plant
E. H_2O moves within the plant via capillary action

64. Angiosperms produce seeds inside protective structures called:

A. petals
B. cones
C. ovaries
D. pollen grains
E. stomata

65. Which correctly relates to an angiosperm structure and what it develops into?

A. embryo → fruit
B. ovule → fruit
C. pollen → grain seed
D. fruit → ovule
E. ovary → fruit

66. A pollen grain landing on a stigma of a flower of the same species produces:

A. flower
B. stamen
C. gametophyte
D. pollen tube
E. root hairs

Copyright © Sterling Education

67. Flowering plants that grow, flower, produce seeds, and die in one year are:

A. biennials
B. perennials
C. dicots
D. annuals
E. monocots

68. Ground tissue is in a plant's:

A. roots and stems only
B. roots, stems, and leaves
C. stems only
D. stems and leaves only
E. roots only

69. A seed plant is held in the ground by its:

A. phloem
B. epidermis
C. stems
D. xylem
E. roots

Any duplication (copies, uploads, PDFs) is illegal.

Notes for active learning

 Copyright © Sterling Education

Practice Questions: Photosynthesis

1. What mechanism is used by C_4 plants to conserve water?

A. developing deep roots
B. performing the Calvin cycle during the day
C. developing water storage in their leaves and stems
D. performing the Calvin cycle early in the morning
E. closing the stomata during heat and dryness

2. What mechanism is used by CAM plants to conserve water?

A. performing the Calvin cycle during the day
B. developing water storage in their leaves and stems
C. including carbon dioxide into RuBP
D. opening their stomata during the night
E. closing their stomata during the night

3. Which of the choices is an autotroph?

A. beech tree
B. octopus
C. zooplankton
D. coral
E. dog

4. Which is NOT a true statement about ATP?

A. ATP provides energy for the mechanical functions of cells
B. Used ATP is discarded by the cell as waste
C. ATP consists of ribose, adenine, and three phosphate groups
D. ADP is produced when ATP releases energy
E. ATP conversion to ADP is an exothermic reaction

5. Photosynthesis has a role in the metabolism of a plant through:

A. breaking down sugars into H_2O and O_2
B. converting H_2O into CO_2
C. consuming CO_2 and synthesizing sugars
D. converting O_2 into cellulose
E. converting O_2 into sugars

Any duplication (copies, uploads, PDFs) is illegal.

6. From the figure, all are parts of an ADP molecule, EXCEPT:

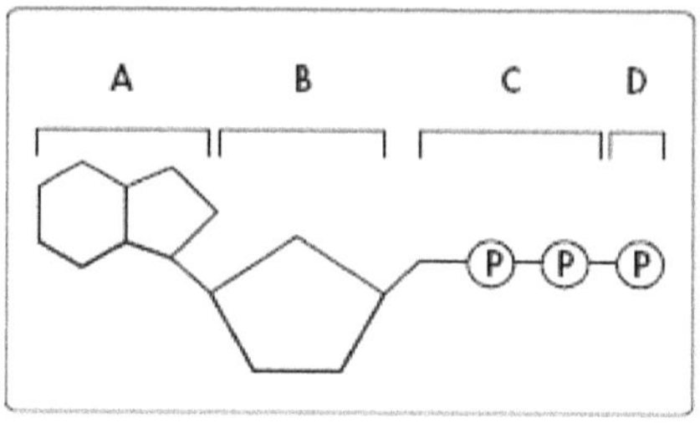

A. structure A
B. structure B
C. structure C
D. structure D
E. structures A and B

7. Within the inner membrane of a chloroplast, stacks of thylakoids are surrounded by:

A. stroma fluid
B. grana
C. chloroplast
D. thylakoids
E. cristae

8. What happens during photosynthesis?

A. Autotrophs consume carbohydrates
B. Autotrophs produce carbohydrates
C. Heterotrophs consume ATP
D. Heterotrophs produce ATP
E. Heterotrophs produce sugars

9. The role of stomata is to facilitate:

A. gas exchange
B. H_2O release
C. Calvin cycle
D. H_2O uptake
E. Chloroplast formation

10. Plants gather the sun's energy using:

A. glucose
B. chloroplasts
C. thylakoids
D. mitochondria
E. pigments

Copyright © Sterling Education

11. Which structures are used by plants to obtain most of their water?

A. Stomata
B. Flowers
C. Roots
D. Chloroplasts
E. Shoots

12. Which structure in the image illustrates a single thylakoid?

A. Structure A
B. Structure B
C. Structure C
D. Structures A and B
E. Structures B and C

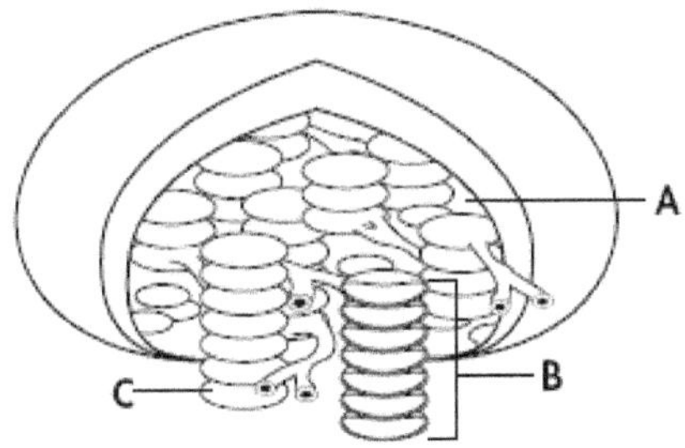

13. Membranous structures in chloroplasts are:

A. organelles
B. grana
C. stomata
D. plasma
E. thylakoids

14. Chlorophyll is within chloroplast in the:

A. thylakoid space
B. thylakoid membrane
C. stroma
D. ATP
E. space between the inner and outer membrane of the thylakoid

15. Grana is/are:

A. stacks of membranous sacs
B. chloroplast pigments
C. fluid in the chloroplasts
D. the space between the inner and outer membrane of chloroplasts
E. the membrane-bound compartment of chloroplasts

Any duplication (copies, uploads, PDFs) is illegal.

16. Which chemical shown in the figure is an electron carrier molecule?

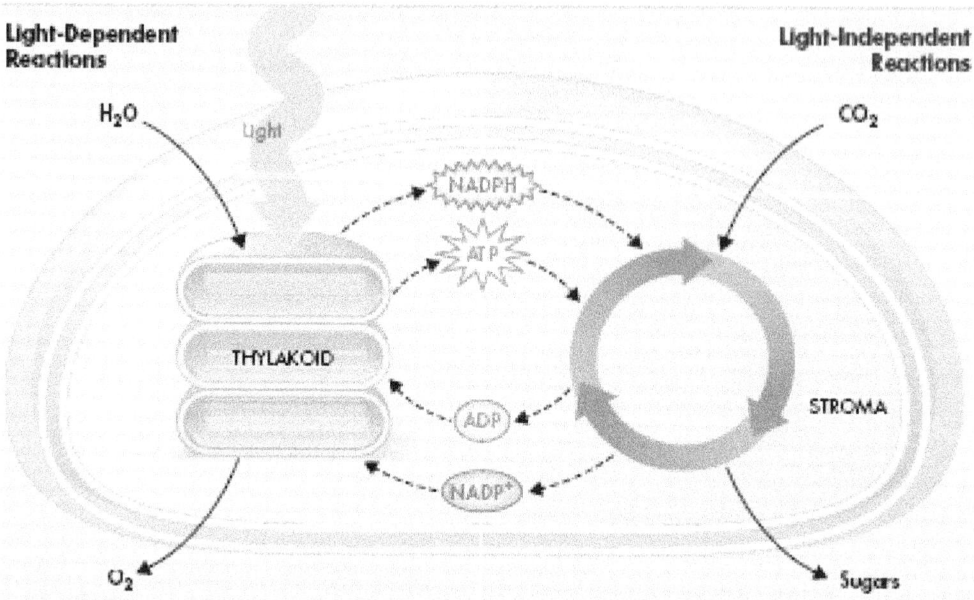

A. $NADP^+$

B. Oxygen

C. H_2O

D. CO_2

E. Sugars

17. Which is likely to result if a shade-tolerant plant receives minimal water while receiving its necessary light?

A. Accelerated plant growth

B. Low ATP production

C. Increased output of oxygen

D. Increased consumption of carbon dioxide

E. Plant death

18. Why are electron carriers needed for transporting electrons from one part of the chloroplast to another?

A. High-energy electrons get their energy from electron carriers

B. High-energy electrons are not soluble in the cytoplasm

C. High-energy electrons are easily oxidized

D. High-energy electrons would be destroyed

E. High-energy electrons are highly reactive

Copyright © Sterling Education

19. Which equation best describes the reaction of photosynthesis?

A. $6\ O_2 + 6\ CO_2 \rightarrow C_6H_{12}O_6 + 6\ H_2O$
B. $C_6H_{12}O_6 + 6\ H_2O \rightarrow 6\ CO_2 + 6\ O_2$
C. $C_6H_{12}O_6 \rightarrow 6\ CO_2 + 6\ H_2O + 6\ O_2$
D. $6\ H_2O + 6\ CO_2 \rightarrow C_6H_{12}O_6 + 6\ O_2$
E. $6\ CO_2 + 6\ O_2 + 6\ H_2O \rightarrow C_6H_{12}O_6$

20. What is the result of an increase in light intensity that a plant receives?

A. The rate of photosynthesis increases and then levels off
B. The rate of photosynthesis does not change
C. The rate of photosynthesis increases indefinitely with light intensity
D. The rate of photosynthesis decreases indefinitely with light intensity
E. The rate of photosynthesis increases indefinitely to the square root of the light intensity

21. In redox reactions of photosynthesis, electron transfer goes as follows:

A. $H_2O \rightarrow CO_2$
B. $C_6H_{12}O_6 \rightarrow O_2$
C. $O_2 \rightarrow C_6H_{12}O_6$
D. $CO_2 \rightarrow H_2O$
E. $C_6H_{12}O_6 \rightarrow H_2O$

22. Which is NOT a step in the light-dependent reactions?

A. ATP synthase allows H^+ ions to pass through the thylakoid membrane
B. ATP and NADPH are used to produce high-energy sugars
C. High-energy electrons move through the electron transport chain
D. Pigments in photosystem II absorb light
E. Electrons move from photosystem II to photosystem I

23. Which is the product of the light reaction of photosynthesis?

A. sugar
B. carbon monoxide
C. water
D. carbon dioxide
E. oxygen

Any duplication (copies, uploads, PDFs) is illegal.

24. Photosystems I and II are in the:

A. cell membrane
B. Calvin cycle
C. thylakoid membrane
D. stroma
E. matrix of the mitochondrion

25. Which components from the light reactions are required by the C_3 cycle?

A. NADH and RuBP
B. NADPH and ATP
C. ATP and NADH
D. glucose and $NADP^+$
E. water and $NADP^+$

26. Which pathway is the correct flow of electrons during photosynthesis?

A. Photosystem I → Calvin cycle → $NADP^+$
B. Light → Photosystem I → Photosystem II
C. H_2O → Photosystem I → Photosystem II
D. O_2 → ADP → Calvin cycle
E. H_2O → $NADP^+$ → Calvin cycle

27. Light-dependent reactions of photosynthesis produce:

A. H_2O and RuBP
B. ATP and CO_2
C. NADH and ATP
D. NADPH and ATP
E. H_2O and O_2

28. The Calvin cycle occurs in the:

A. chlorophyll molecules
B. thylakoid membranes
C. photosystems
D. stroma
E. matrix

29. A wavelength of which light is reflected by a blue-colored plant?

A. violet
B. yellow
C. blue
D. lime
E. purple

Copyright © Sterling Education

30. What are the three parts of an ATP molecule?

A. Adenine, ribose, and three phosphate groups
B. NADH, NADPH, and $FADH_2$
C. Adenine, thylakoid, and a phosphate group
D. Stroma, grana, and chlorophyll
E. Adenine, ribose, and a phosphate group

31. When observing a plant, its visible color is:

A. wavelength emitted by the plant
B. wavelength from the return of excited electrons to their ground state
C. wavelength being reflected by that plant
D. wavelength absorbed by the pigment of the plant
E. wavelength from the excited electrons of the plant

32. Energy is released from ATP when:

A. a phosphate group is removed
B. ATP is exposed to sunlight
C. adenine binds to ribose
D. a phosphate group is added
E. a base is removed

33. The wavelength energy least utilized by photosynthesis is:

A. orange
B. blue
C. violet
D. cyan
E. green

34. Organisms, such as plants, which make their food are:

A. symbiotic
B. parasitic
C. autotrophs
D. heterotrophs
E. omnivores

35. During the fall foliage, which pigments are responsible for the yellow, red, and orange colors of leaves?

A. carotenoids
B. chlorophyll *b*
C. melanin
D. chlorophyll *a*
E. anthocyanin

Any duplication (copies, uploads, PDFs) is illegal.

36. Which organism is a heterotroph?

A. sunflower
B. flowering plant
C. alga
D. mushroom
E. deciduous trees

37. In addition to chlorophyll *a*, plants have chlorophyll *b* and carotenoids accessory pigments because:

A. there is not enough chlorophyll *a* produced for the plant's energy needs
B. plants must have leaves of different colors
C. these pigments reflect more energy
D. these pigments protect plants from UV radiation
E. these pigments absorb energy wavelengths that chlorophyll *a* does not

38. Plants get the energy they need for photosynthesis by absorbing:

A. chlorophyll *b*
B. energy from the sun
C. high-energy sugars
D. chlorophyll *a*
E. both chlorophyll *a* and *b*

39. A discrete packet of light is a(n):

A. photon
B. neutron
C. proton
D. wavelength
E. electron

40. Most plants appear green because chlorophyll:

A. does not absorb violet light
B. does not absorb green light
C. absorbs violet light
D. absorbs green light
E. generate wavelengths of green light

41. In visible light, shorter wavelengths carry:

A. more energy
B. more photons
C. more red color
D. less energy
E. fewer photons

Copyright © Sterling Education

42. Interconnected sacs of a membrane suspended in a thick fluid are:

A. stroma
B. thylakoids
C. chlorophyll
D. grana
E. stomata

43. When a molecule absorbs a photon, one of its electrons is raised to a(n):

A. higher state
B. excited state
C. lower state
D. ground state
E. valence

44. What is the function of $NADP^+$ in photosynthesis?

A. Photosystem
B. Pigment
C. Electron carrier
D. High-energy sugar
E. Absorb photons

45. A molecule releases energy gained from the absorption of a photon through:

I. fluorescence II. heat III. light

A. I only
B. II only
C. III only
D. II and III only
E. I, II and III

46. Photosynthesis uses sunlight to convert water and carbon dioxide into:

A. oxygen and high-energy sugars
B. ATP and oxygen
C. high-energy sugars and proteins
D. oxygen and carbon
E. ATP and high-energy sugars

47. Photosystems are in:

A. grana
B. stroma
C. stomata
D. thylakoid membranes
E. matrix of mitochondria

Any duplication (copies, uploads, PDFs) is illegal.

48. The light-dependent reactions take place:

A. within the thylakoid membranes
B. in the outer membrane of the chloroplast
C. in the stroma of the chloroplast
D. within the mitochondria membranes
E. in the inner membrane of the chloroplast

49. Which compound is in the photosystem's reaction center?

A. FADPH
B. rhodopsin
C. ATP
D. ADP
E. chlorophyll *a*

50. What are the products of light-dependent reactions?

A. CO_2 gas, O_2 gas, and NADPH
B. ATP, CO_2 gas, and NADPH
C. ATP, NADPH, and O_2 gas
D. O_2 gas and glucose
E. ATP, NADPH, and glucose

51. Which is the correct matching of molecules and products in the Calvin cycle?

A. ATP + NADPH + carbon dioxide ⇒ sugar
B. light + water + carbon dioxide ⇒ sugar and oxygen
C. ATP + NADPH + carbon dioxide ⇒ sugar and oxygen
D. light + water + carbon dioxide ⇒ sugar
E. carbon dioxide + light ⇒ oxygen and water

52. Which activities happen within the stroma?

A. Electrons move through the electron transport chain
B. The Calvin cycle produces sugars
C. ATP synthase produces ATP
D. Photosystem I absorbs light
E. Photosystem II produces sugar

Copyright © Sterling Education

53. Oxygen released by a photosystem comes from:

A. ATP
B. carbon dioxide
C. Chlorophyll *a*
D. water
E. light

54. The Calvin cycle is another name for the:

A. photosynthesis reaction
B. electron transport chain
C. light-independent reactions
D. light-dependent reactions
E. Photosystem I

55. Electrons for light reactions originate from:

A. NADPH
B. water
C. light
D. carbon dioxide
E. ATP

56. Which mechanism occurs in light reactions of photosynthesis and cellular respiration?

A. C_3 cycle
B. Beta oxidation
C. Glycolysis
D. Krebs cycle
E. Electron transport chain

57. During photosynthesis, an H^+ ion gradient is formed across:

A. mitochondrial inner membrane
B. inner chloroplast membrane
C. thylakoid membrane
D. mitochondrial outer membrane
E. stroma membrane

Any duplication (copies, uploads, PDFs) is illegal.

Notes or active learning

Copyright © Sterling Education

Detailed Explanations: Plants

Answer Key

1: E	11: C	21: A	31: D	41: D	51: C	61: A
2: A	12: B	22: B	32: E	42: E	52: A	62: B
3: B	13: E	23: E	33: B	43: A	53: E	63: A
4: A	14: D	24: D	34: A	44: C	54: B	64: C
5: E	15: C	25: C	35: D	45: E	55: D	65: E
6: C	16: A	26: E	36: C	46: C	56: B	66: D
7: D	17: D	27: A	37: A	47: B	57: A	67: D
8: C	18: E	28: C	38: E	48: A	58: E	68: B
9: B	19: A	29: B	39: C	49: D	59: C	69: E
10: D	20: B	30: D	40: B	50: B	60: D	

1. E is correct.

Angiosperms produce pollen grains in the *anther*.

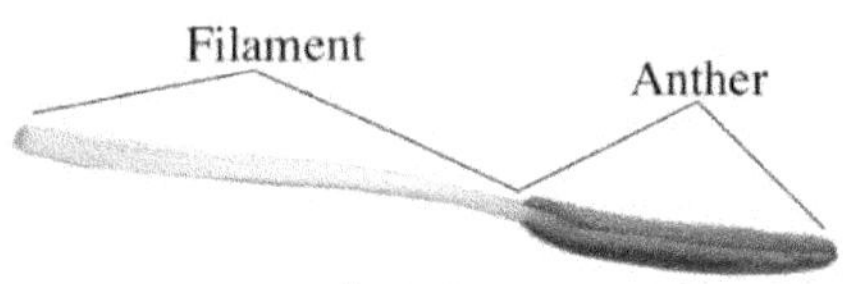

stamen

Stamen is a flower pollen-producing reproductive organ.

Stamen typically has a *filament* stalk and an *anther* with microsporangia.

Anthers are part of the stamen, often two-lobed, and attached to the filament at the base or in the middle portion.

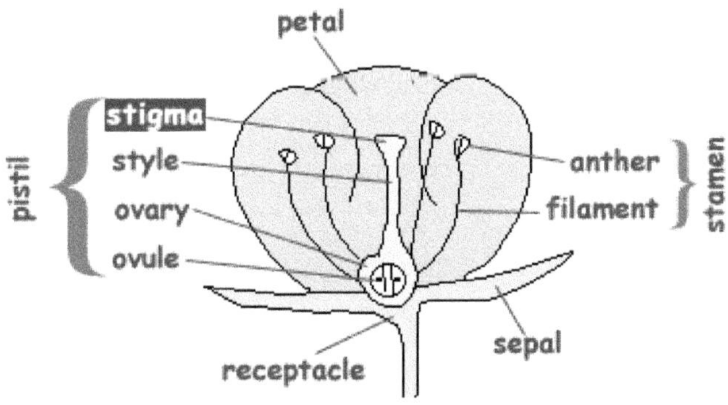

Anatomy of an angiosperm flower with female pistil and male stamen

A: *ovule* is the structure that gives rise to and contains female reproductive cells with the megaspore-derived female *gametophyte* (*megagametophyte*) in the center.

continued...

Megagametophyte (i.e., embryo sac in flowering plants) produces egg (or several) cells for fertilization.

Ovule develops into a seed after fertilization.

Ovary is the enlarged basal portion of the pistil where *ovules* are produced.

B: *sepals* are a part of the flower of *angiosperms* (i.e., flowering plants), and morphologically, sepals and petals are modified leaves.

C: *stigma* is part of the pistil where pollen germinates.

D: *carpels* are (categorically) appendages containing and enclosing ovules.

2. A is correct.

Leaves are plant organs with a primary function of *photosynthesis*.

B: *flowers* have the sole function of sexual reproduction. They are often the most prominent part of a plant.

Flowers' beauty and fragrance evolved pollinators (e.g., insects or birds), which are essential for reproduction. Flowers also help classify plants.

C: *root* is part of a vascular plant, usually underground. Its functions are anchorage of the plant, absorption of water and dissolved minerals and conduction of these nutrients to the stem, and storage of reserve foods.

D: *stems* primary functions are to support the leaves; conduct water and minerals to the leaves, where they can be converted into usable products by photosynthesis; and transport these products from the leaves to other parts of the plant, including the roots.

E: *stomata* are composed of a pair of specialized epidermal cells known as *guard cells*. They regulate gas exchange (e.g., CO_2) and control water loss by changing the size of the stomatal pore.

3. B is correct.

Most plants get the water they need from the soil. *Roots* absorb the bulk of water uptake by most plants.

Plants can absorb water through leaves, but it is not efficient. Plants can take in some of that surface water if water condenses on the leaf during high humidity (e.g., fog).

4. A is correct.

In angiosperms, *xylem* consists of *tracheids* and *vessel elements*.

continued...

 Copyright © Sterling Education

Tracheids and *vessel elements* are elongated cells in vascular plant xylem that transport water and mineral salts.

Tracheids are elongated cells transporting water and mineral salts through the xylem of vascular plants.

Tracheids provide *physical support* and *conduct water and dissolved minerals upwards* and are the only such elements in conifers and ferns.

Tracheids, unlike vessel elements, do not have perforations.

B: *parenchyma* is a type of simple permanent tissue and a significant part of ground tissues in plants, where other tissues, like vascular tissues, are embedded.

Parenchyma is *non-vascular*, with *undifferentiated cells* modified to perform various functions.

C: *sieve tube elements* are specialized cells essential for the phloem's function, a highly organized tissue that transports organic compounds from photosynthesis.

D: *albuminous cells* have a similar role to companion cells but are associated with sieve cells only and are only in seedless vascular plants and gymnosperms.

E: *companion cells* perform the cellular functions of a sieve-tube element in nucleated plant cells. They usually have a larger number of ribosomes and mitochondria.

Companion cell cytoplasm is connected to the sieve-tube element by *plasmodesmata* (i.e., gated plant cell wall channels that allow movement of molecules between cells).

5. E is correct.

Flowers in angiosperms have the sole function of sexual reproduction and are often the most prominent part of a plant. Flowers that animals pollinate do not exist as small structures.

Flowers' beauty and fragrance evolved pollinators (e.g., insects or birds), which are central to the reproductive process. Flowers also help classify plants.

6. C is correct.

Without gas exchange, a plant could not make food (i.e., carbohydrates such as glucose).

Stomata (stoma, singular) and *lenticels* regulate gas exchange in plants.

Plant epidermis has stomata, tiny pores that control transpiration and gas exchange with air. During the day, photosynthesis occurs, carbon dioxide (CO_2) is consumed, and oxygen (O_2) is released into the atmosphere.

Without gas exchange, plants cannot complete photosynthesis without carbon dioxide from the atmosphere.

Any duplication (copies, uploads, PDFs) is illegal.

Photosynthesis

$6CO_2 + 6H_2O \longrightarrow C_6H_{12}O_6 + 6O_2$

carbon dioxide + water → glucose + oxygen

Photosynthesis reaction:

$$6\ CO_2 + 6\ H_2O \rightarrow C_6H_{12}O_6 + 6\ O_2$$

Six carbon dioxide (CO_2) and six water (H_2O) molecules are converted by light energy (photons from the sun) captured by the photopigment chlorophyll into a sugar molecule (e.g., $C_6H_{12}O_6$ as glucose), and six expired oxygen (O_2) molecules.

7. D is correct.

Filament and *anther* comprise a stamen as a male reproductive structure of flowering plants.

Stamen (male organ) consists of an *anther* (site of pollen development) and, in most species, a stalk-like *filament*, which transmits water and nutrients to the anther and positions it for pollen dispersal.

Pistil (female organ) is the *ovule-producing* organ of a flower with a stigma, style, and ovule.

Ovary often supports a long *style*, topped by a *stigma.*

Fruit is a mature ovary, and the *mature ovule* is a *seed.*

Stigma is part of the pistil where *pollen germinates*.

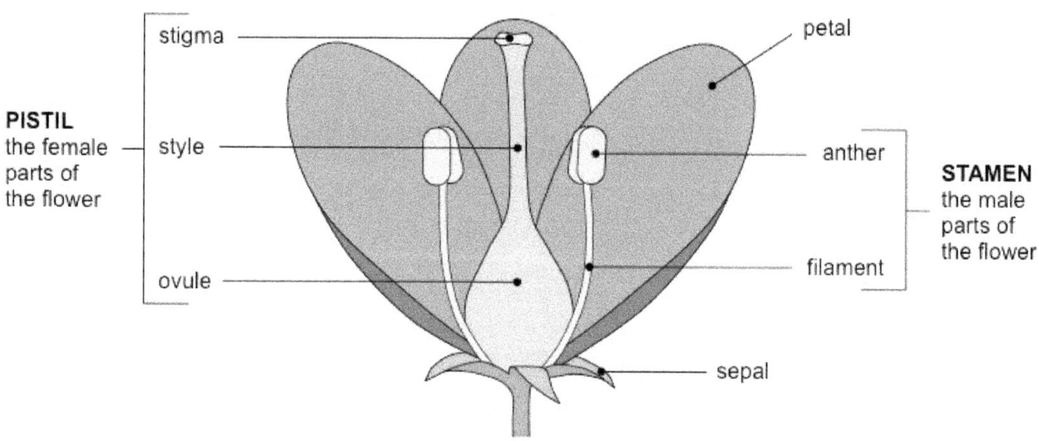

Pistil (stigma, style, and ovule) is female, and stamen (anther and filament) is the male part of the flower

Copyright © Sterling Education

8. C is correct.

Plants living on land must *conserve water*.

Water loss was the primary challenge for plants moving from aquatic to terrestrial environments.

Cutin, stomata, roots, and root hairs help terrestrial plants absorb and conserve water.

Roots and root hairs allow plants to absorb nutrients and water from the soil.

Plants have adapted to environments to sustain water balance (e.g., closing of stomata), as, without it, all processes stop, and the plant would not be able to survive.

For example, when the humidity is low, water is more likely to evaporate quickly from the leaf surface, and plants often close or partially *close their stomata* to maintain a stable water balance in the leaf.

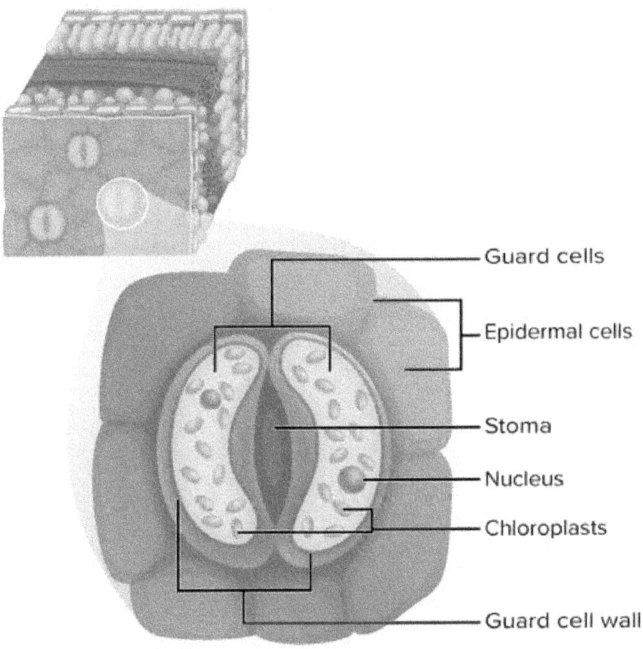

Leaf underside cross-section with an open stoma, guard cells, and support structures

9. B is correct.

Vascular tissue in plants consists of *xylem* and *phloem*.

Xylem is in vascular bundles (non-woody plants and non-woody parts of woody plants), secondary xylem from vascular cambium meristem in woody plants, and non-bundles (e.g., many ferns).

continued...

Any duplication (copies, uploads, PDFs) is illegal.

Xylem transports water, soluble minerals, and nutrients from the roots throughout the plant and replaces water lost during transpiration (water movement upwards) and photosynthesis (sugars formed in leaves).

Phloem is living tissue in vascular plants that transports (i.e., translocation) organic nutrients (e.g., fructose, sucrose) synthesized during photosynthesis to vascular plant parts.

Phloem tissue consists of *conducting cells* (i.e., sieve elements), *parenchyma cells* (specialized companion and albuminous cells and unspecialized cells), and *supportive cells* (i.e., fibers and sclereids).

10. D is correct.

Vascular tissue in plants consists of *xylem* and *phloem*.

Xylem is in vascular bundles (non-woody plants and non-woody parts of woody plants), secondary xylem from vascular cambium meristem in woody plants, and non-bundles (e.g., many ferns).

Xylem transports water, soluble minerals, and nutrients from roots and replaces water lost during photosynthesis.

Phloem is living tissue in vascular plants that transports downward (*translocation*) organic nutrients (e.g., sucrose) synthesized during photosynthesis to vascular plant parts.

Phloem tissue consists of *conducting cells* (i.e., sieve elements), *parenchyma cells* (specialized companion and albuminous cells and unspecialized cells), and *supportive cells* (i.e., fibers and sclereids). For example, the phloem is the innermost layer of bark in trees.

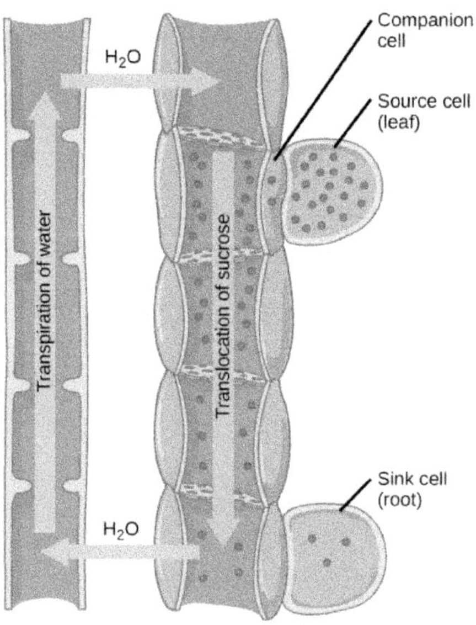

Xylem (on the left) functions for the transpiration of H_2O from roots, with phloem (on right) for the translocation of sugars from leaves during photosynthesis

 Copyright © Sterling Education

A: *epidermis* (or d*ermal tissue*) is the outer covering of a plant (roots, stems, leaves, flowers, fruits, and seeds); usually, one cell layer thick, and its cells lack chloroplasts.

B: *parenchyma* is the most versatile and abundant cell of the three types of ground tissue.

Sclerenchyma is dead ground tissue that supports thick (primary and secondary) cell walls.

Collenchyma is living ground tissue that supports living tissue with irregular walls. They have thick deposits of *cellulose* in cell walls and appear *polygonal* in cross-section.

C: *meristem* is tissue in most plants containing undifferentiated cells in growth zones that give rise to plant organs and keep the plant growing.

E: *rhizoid* is a structure in plants and fungi functioning like a root for *support or absorption.*

11. C is correct.

Ancestors of land plants likely evolved from *streptophyte algae* (freshwater green algae) 500 million years ago.

Likely, plants evolved from living in water to inhabiting land because of genes they took up from bacteria.

12. B is correct.

Plants reproduce vegetatively by mitosis, whereby a single parent produces cloned offspring; no seeds or spores are generated, and does not involve the fusion of gametes.

Vegetative reproduction is *asexual* reproduction in plants when a new daughter plant grows from a fragment of the parent plant or grows from a *specialized reproductive structure* (e.g., stolon, rhizome, tuber, corm, or bulb).

Vegetative part (i.e., root, stem, leaf) of the parent is preserved and propagated to generate offspring.

13. E is correct.

Plants relying on animals for dispersion have *seeds enclosed in fleshy fruits*, encouraging animals to eat them.

Animal digests fruit flesh, but the seed enclosed in a thick indigestible coating passes through an animal's digestive tract without being digested.

When an animal excretes undigested seeds, they often land in a place distant from their parent and can germinate when conditions are right.

Any duplication (copies, uploads, PDFs) is illegal.

14. D is correct.

Plant shoots consist of any plant stem with appendages, leaves and lateral buds, flowering stems, and flower buds that can regenerate the plant.

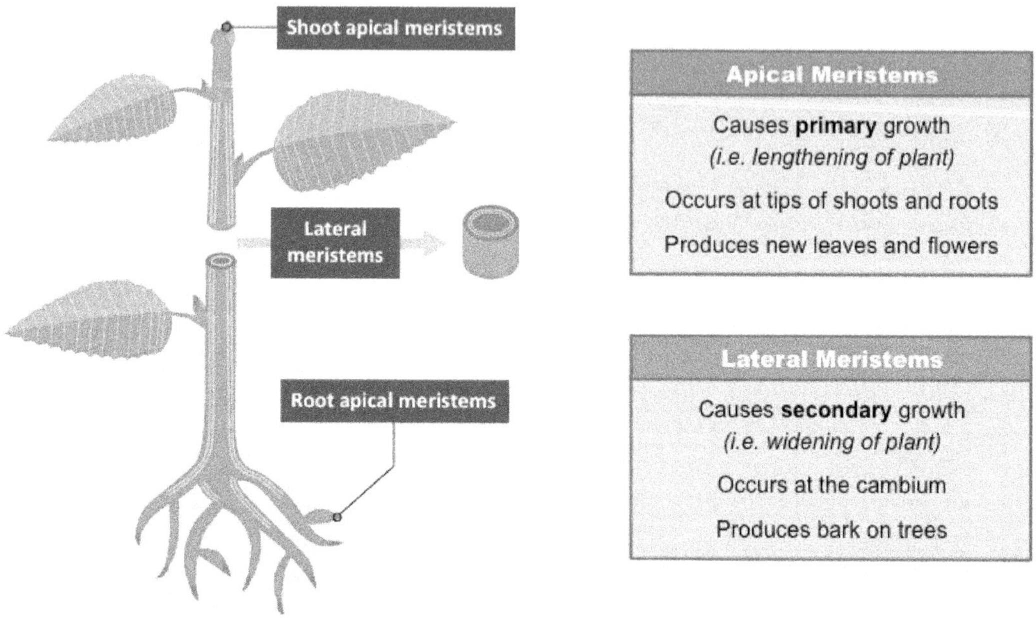

Apical (shoots and roots) and lateral (cambium) meristems are regions of mitotic cell growth

Meristematic cells (i.e., centers of mitotic cell division) at the tip of roots render desirable and suitable raw material to study the stages of mitosis.

Meristems mitotic cells are undifferentiated *self-renewing stem cells* from which most structures arise.

Apical meristems cause primary growth to elongate the plant (e.g., shoot growth)

Lateral meristems cause secondary growth to widen the plant (e.g., growth rings of trees)

Meristematic cells are also responsible for keeping the plant growing.

15. C is correct.

Animals would not likely live on land if plants did not migrate to land.

Land could not be colonized by other organisms until land plants became established because plants became a food source for many land animals.

Copyright © Sterling Education

16. A is correct.

Xylem and *phloem* are the two types of transport tissue in vascular plants.

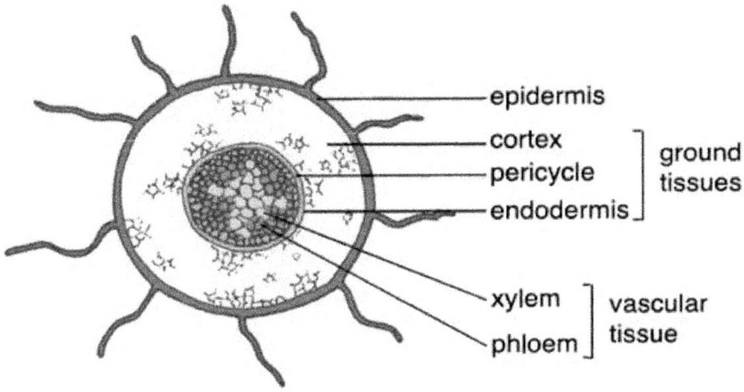

Root cross-section with xylem and phloem vascular tissue

Xylem is in vascular bundles (non-woody plants and non-woody parts of woody plants), secondary xylem laid down by vascular cambium (meristem) in woody plants, and not divided into bundles (e.g., many ferns).

Xylem transports water, soluble minerals, and nutrients from roots and replaces water lost during photosynthesis.

Phloem is the tissue that transports (i.e., translocation) soluble organic nutrients synthesized during photosynthesis (e.g., glucose) to all vascular plant parts where needed; it is the innermost layer of tree bark.

Phloem tissue is *conducting cells* (i.e., sieve elements), *parenchyma cells* (specialized companion and albuminous cells and unspecialized cells), and *supportive cells* (i.e., fibers and sclereids).

D: *meristem* is the tissue in most plants that contains undifferentiated cells in the growth zones of plants. Meristematic cells give rise to various plant organs and keep the plant growing.

17. D is correct.

Vegetative reproduction is asexual reproduction in plants when a daughter plant grows from a fragment of the parent or grows from a specialized reproductive structure (e.g., stolon, rhizome, tuber, corm, or bulb).

Plants reproduce vegetatively by mitosis, whereby a single parent produces *identical cloned offspring*; no seeds or spores are generated, and do not involve genetic diversity with the fusion of gametes.

continued...

Any duplication (copies, uploads, PDFs) is illegal.

Offspring are produced by mitosis and lack genetic recombination generated by meiosis.

Progeny are identical clones of the parent and lack genetic diversity to survive a shift in the physical environment.

18. E is correct.

Endodermis is the layers of cells enclosing the vascular tissue (xylem and phloem) in a root's central region.

A: *apical meristem* is meristematic tissue at the tips of stems and roots, enabling an extension in length.

B: *epidermis* (or d*ermal tissue*) is the outer covering of a plant (roots, stems, leaves, flowers, fruits, and seeds); usually, one cell layer thick, and its cells lack chloroplasts.

C: *cortex* (woody plants) is between the *periderm* (bark) and *vascular tissue* (phloem) that transports materials into the central cylinder of the root through *diffusion.*

Cortex may be used to store food in the form of *starch.*

D: *meristems* are undifferentiated self-renewing stem cells from which most structures arise.

Apical meristems cause primary growth to elongate the plant (e.g., shoot growth)

Lateral meristems cause secondary growth to widen the plant (e.g., growth rings of trees)

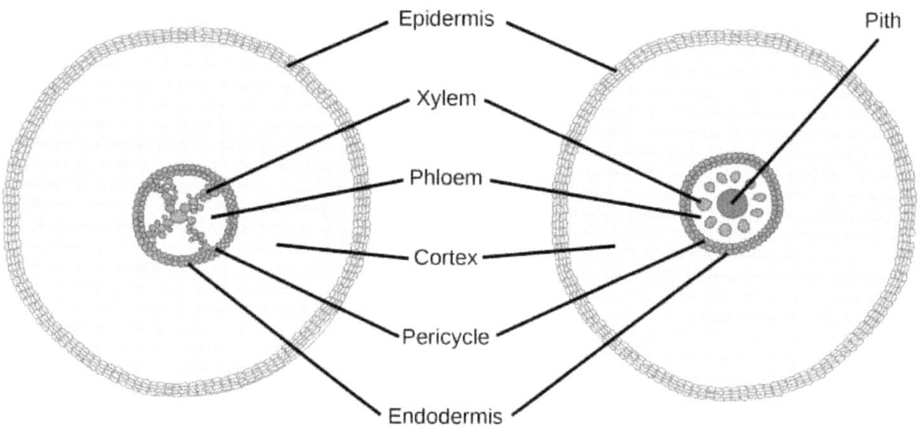

Epidermis, vascular tissue, cortex, and endodermis for dicot (left) and monocot root (right)

Copyright © Sterling Education

19. A is correct.

Ploidy number is the fundamental difference between a sporophyte and a gametophyte.

Gametophyte is haploid (1N)

Gametophyte is the sexual phase with haploid (1N) male and female organs (*gametangia*) developing and producing eggs and sperm gametes through mitosis as sexual reproduction.

Egg and sperm unite in fertilization, forming a 2N zygote (fertilized egg) that develops into a diploid sporophyte, which produces unicellular spores through meiosis.

These develop a new gametophyte phase.

Gametophytes mature within spores, not free-living, as gametophytes of other seedless vascular plants.

Sporophyte is diploid (2N)

Sporophyte is the nonsexual phase when a diploid (2N) plant body grows and produces spores through meiosis.

These spores divide mitotically to produce haploid gametophytes.

Spore release in a suitable environment leads to germination and a new generation of gametophytes.

Completing the life cycle requires water, as male gametes, propelled by flagella, swim to the female gametes.

After fertilization, the zygote matures and grows into a *sporophyte*, forming sporangia (or *spore vessels*) in which mother cells undergo meiosis and produce haploid spores.

Heterosporous seedless plants are seen as the evolutionary forerunners of seed plants.

Seed plants (or *spermatophytes*) form two types of spores:

Megaspores (female) develop into gametophytes that produce eggs.

Microspores (male) mature into gametophytes that generate sperm.

Alternations of generations are distinct phases in the life history of organisms, each phase producing the other.

Gametophytes (haploid) represent the sexual phase in the alteration of generations.

Sporophyte (diploid) is the dominant and nonsexual phase in the life cycle.

continued...

Any duplication (copies, uploads, PDFs) is illegal.

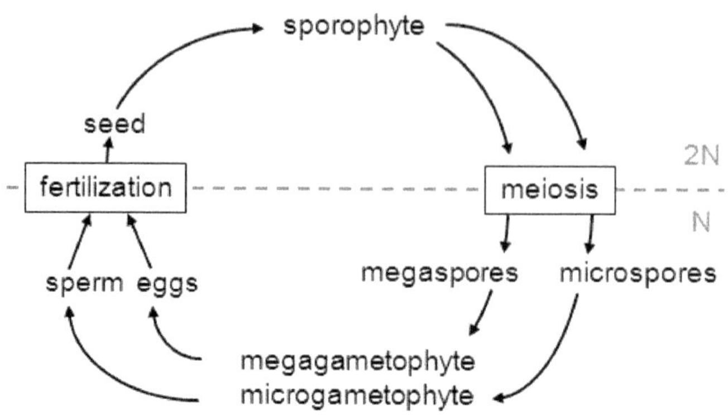

Alternation of generations with sporophyte is diploid (2N), while a gametophyte is haploid (1N) in the life cycle

Two gametes join during fertilization as a diploid zygote, dividing mitotically to form a new sporophyte.

Sporophytes	**Gametophytes**
Sporophytes undergo *meiosis*	Gametophytes undergo *mitosis*
Form *spores*	Form *gametes*
Diploid (2N) plants	Haploid (1N) plants
Two sets of chromosomes	Single set of chromosomes
Reproduce *asexually*	Produce *sexually*
Diploid spore cells undergo meiosis, forming haploid (1N) meiospores	Gametes fuse (or undergo fertilization), forming diploid (2N) zygotes

Gametophytes and sporophytes with spore versus gamete distinctions

Bryophytes and pterophytes characterize the alternation of generations.

Pollen and seed allowed seed plants to break their dependence on water for reproduction and development of the embryo and thrive on dry land.

Pollen grains carry male gametes with small haploid (1N) cells encased in a protective coat to prevent desiccation (i.e., drying) and mechanical damage.

continued...

 Copyright © Sterling Education

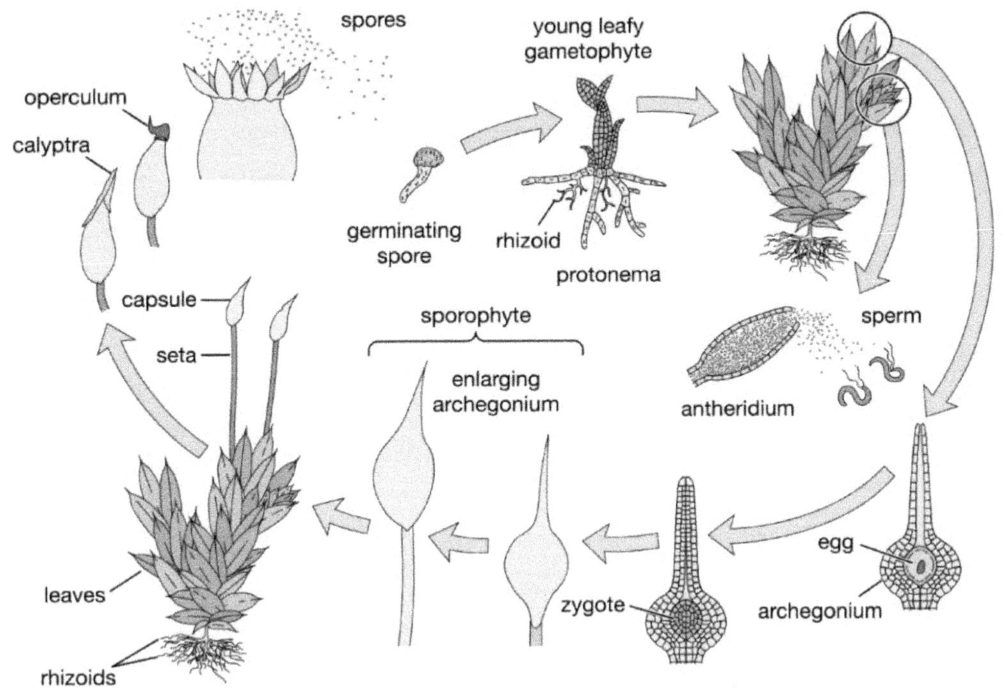

Alternations of generations for gametophytes and sporophytes in mosses

20. B is correct.

Dermal tissue (epidermis) is the outer covering of a plant (e.g., roots, stems, leaves, flowers, fruits, and seeds), usually one cell layer thick, with cells lacking chloroplasts.

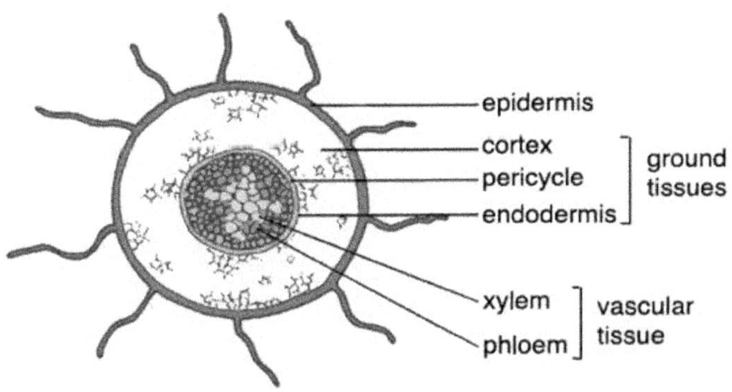

Root cross-section with xylem and phloem vascular tissue

21. A is correct.

Vascular plant gametophytes tend to be much smaller than sporophytes, sometimes only a few cells in size.

continued...

Gametophytes have become smaller, while sporophytes have become larger as plants evolved.

For example, a pollen grain is a male gametophyte in vascular plants.

In seed plants, evolutionary trends led to a dominant sporophyte generation with a reduction in gametophyte size from a conspicuous structure to a microscopic cluster of cells enclosed with sporophyte tissue.

> *Gametophyte* is the sexual haploid phase in the life cycle when a haploid (1N) organism develops from a haploid (1N) spore with one set of chromosomes.
>
> *Sporophyte* is the diploid stage in the life cycle and develops from a diploid (2N) zygote produced when a haploid (1N egg) cell is fertilized by a 1N sperm.
>
> Sporophyte cells have a set of chromosomes, one set from each parent.

Diploid plants with alternation of generations begin from a diploid (2N) zygote and produce haploid (1N) spores by meiotic division.

Seeds and pollen (adaptations to drought) distinguish seed plants from seedless vascular plants.

Both adaptations were critical to the colonization of land.

Pollen can travel far from the sporophyte that bore it, spreading the plant's genes and reducing competition with genetically similar plants.

Seed offers embryo protection, nourishment, and a mechanism to maintain dormancy for thousands of years, surviving harsh environments and germinating when growth conditions are optimal.

Seeds disperse through space and time; evolution favored seed plants as successful and familiar plants.

22. B is correct.

Dormant seeds are ripe seeds that do not germinate because they are subject to external environmental conditions that prevent the initiation of metabolic processes and cell growth.

Germination is when an organism grows from a seed or spore. It applies to the sprouting of a seedling from a seed of an angiosperm (flowering plant) or gymnosperm (seed-producing plant) and the growth of the pollen tube from the pollen grain of a seed plant.

When conditions (e.g., water, light) become favorable, the dormant seed germinates, and the embryonic tissues resume growth, developing into a seedling.

 Copyright © Sterling Education

23. E is correct.

Seeds and fruits result from fertilization or sexual reproduction in plants.

Seeds are formed in gymnosperms and angiosperms. In angiosperms (flowering plants), fruits are matured ovaries that contain seeds.

Ovaries in angiosperms develop into fruit, whereas *ovules* become *seeds* enclosed within the fruit.

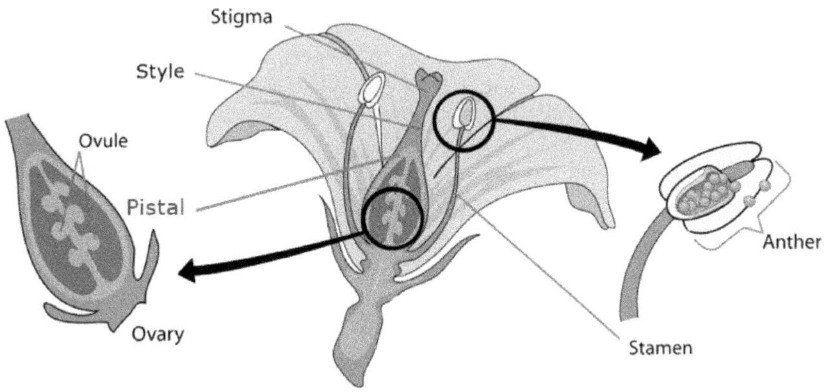

Angiosperm sexual reproduction anatomy with female ovary and male stamen (anther and filament)

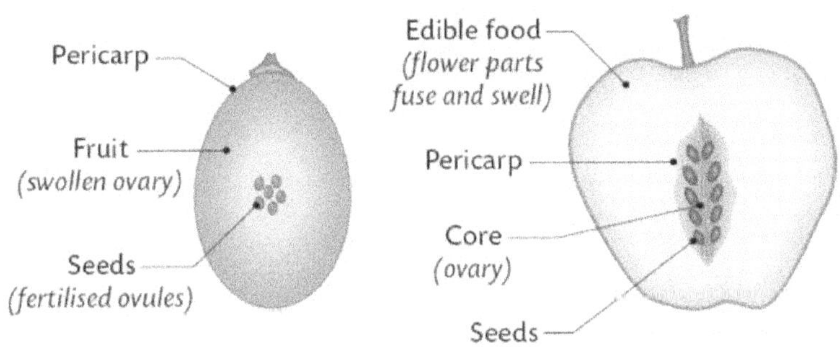

Two types of fruits with grapes as true fruit and apples as false fruit (on right)

True fruit forms from all tissues derived from a ripened ovary and its contents.

False fruit forms from other parts of the plant and the ovary, especially the receptacle.

Receptacle is the part of a flower stalk where parts of the flower are attached.

Sepal is the outer parts of the flower (often green and leaf-like) that enclose a developing bud.

Any duplication (copies, uploads, PDFs) is illegal.

24. D is correct.

Green algae (unicellular or multicellular) are classified as plants because they contain *chloroplasts* and produce food (e.g., carbohydrates such as sucrose) through *photosynthesis*.

Green algae lack many structures of plants.

For example, algae do not have roots, stems, or leaves.

25. C is correct.

Rhizoid is like a root in bryophytes (i.e., mosses, liverworts, and hornworts).

Rhizoids are simple hair-like protuberances extending from bryophyte lower epidermal cells similar in structure and function to *root hairs of vascular land plants*.

Rhizoids form from single cells, whereas roots are multicellular organs composed of multiple tissues that collectively perform a common function.

26. E is correct.

Seeds dispersed to an area far away from the parent plant have less competition with parent plants.

Seeds can be transported (e.g., via animals) to a new location or survive
adverse climate conditions.

Seed offers embryo protection, nourishment, and a mechanism to maintain dormancy for thousands of years, surviving harsh environments and germinating when growth conditions are optimal.

Seeds disperse through space and time; evolution favored seed plants as successful and familiar plants.

27. A is correct.

Root pressure is produced in the vascular cylinder by *osmotic pressure* within cells of a root system that causes sap to rise through a plant stem to the leaves.

Osmotic pressure is the minimum pressure that must be applied to a solution to stop the flow of solvent molecules through a semipermeable membrane (osmosis).

Osmotic pressure is a colligative property and depends on the solution's concentration of solute particles.

Root pressure occurs in the *xylem* of some vascular plants when the soil moisture level is high (e.g., at night or when transpiration is low during the day).

continued...

Copyright © Sterling Education

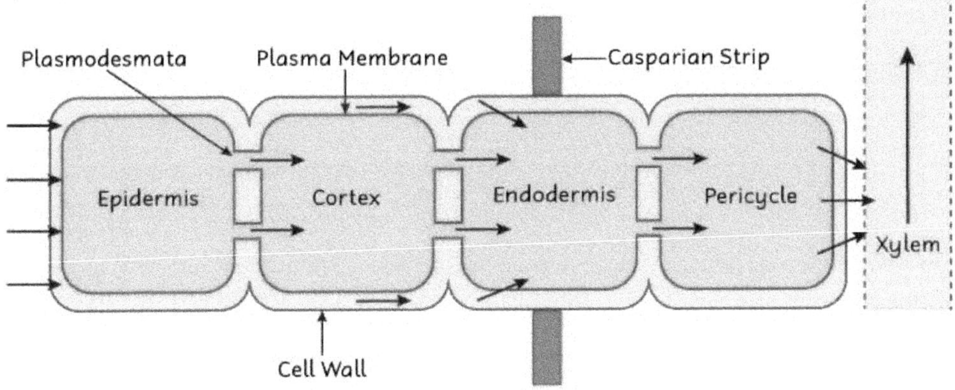

Root pressure in vascular plants is produced by osmotic pressure

Diffusion and *osmosis* (movement of water) use passive transport and do *not require energy* input.

In diffusion and osmosis, particles move from an area of *higher concentration* to one of *lower concentration.*

Casparian strips form a barrier to apoplastic flux (i.e., through cell walls), forcing ions through the selectively permeable plasma membrane into the cytoplasm rather than moving along cell walls.

28. C is correct.

Evidence suggests that green-blue algae (protists) were the first plants to evolve from a group of green algae 850 mya, and algae-like plants might have evolved as early as 1 billion years ago.

Green algae are classified as plants because they contain *chloroplasts* and produce food through *photosynthesis*.

Photosynthesis

$$6CO_2 + 6H_2O \longrightarrow C_6H_{12}O_6 + 6O_2$$

carbon dioxide + water ⟶ glucose + oxygen

Green algae and plants have cellulose in their cell walls and share many photopigments for color.

However, they lack many other structures of true plants.

For example, algae do *not* have roots, stems, or leaves.

Any duplication (copies, uploads, PDFs) is illegal.

29. B is correct.

Germination is the growth of an embryonic plant within a seed, resulting in the formation of a seedling.

Seeds of vascular plants are small packages with fruits (or cones) from the union of male and female gametes.

Fully developed seeds contain an embryo (most plants also store nutrient reserves) encased in a seed coat.

Some plants produce varying numbers of seeds that lack embryos; these are empty seeds that do not germinate.

Most seeds undergo a period of dormancy when there is no active growth.

Seeds can be transported (e.g., via animals) to a new location or survive adverse climate conditions until favorable circumstances develop.

30. D is correct.

Dormancy allows seeds to *germinate* under ideal living conditions (e.g., temperature, humidity). Dormancy is when there is no germination in seeds or tubers even though the required conditions (temperature, humidity, oxygen, and light) exist.

Dormancy uses *hard seed coat impermeability* or *lack of enzymes* (internal dormancy) needed for germination.

31. D is correct.

Primary functions of stems:

1) support the leaves,

2) conduct water and minerals to the leaves for photosynthesis, and

3) transport carbohydrate products from leaves to other plant parts, including roots.

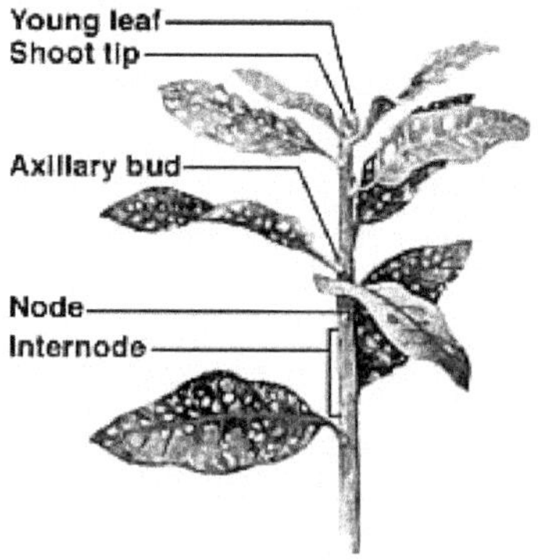

Stems (with meristem growth regions shown) function for support, conduction, and growth

 Copyright © Sterling Education

32. E is correct.

Gibberellins (GAs) are a group of plant hormones that stimulate shoot elongation, seed germination, and fruit and flower maturation.

For example, maturing grapes are treated with gibberellins to promote fruit size.

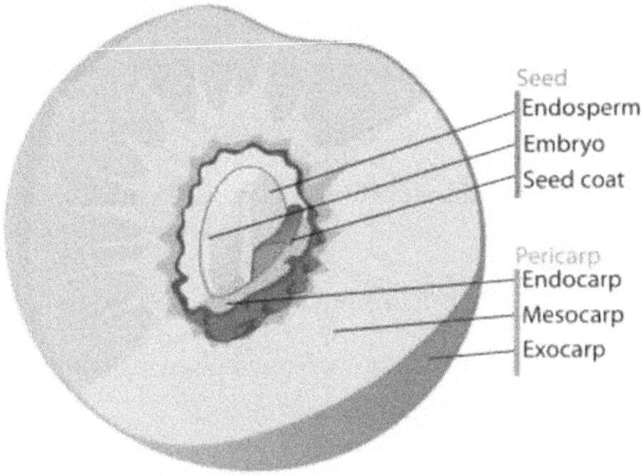

Flowering plant fruit anatomy with seed and pericarp components

Fruit growth is mainly dependent on *auxin* and *cytokinin.*

Seeds are a rich source of hormones, particularly auxins, GA, and cytokinin, which stimulate the growth of surrounding tissues and determine fruit size.

Auxins (the first plant hormones discovered) function primarily in stem elongation by promoting cell growth and are a major growth factor in plants.

Auxins' roles in plants depend on where they are produced.

Fruit ripening uses *ethylene* for texture changes, softening, color, and other ripening processes.

Ethylene level in under-ripe fruit is very low. For example, as the fruit develops, it produces larger amounts which speed up ripening or stage of ripening.

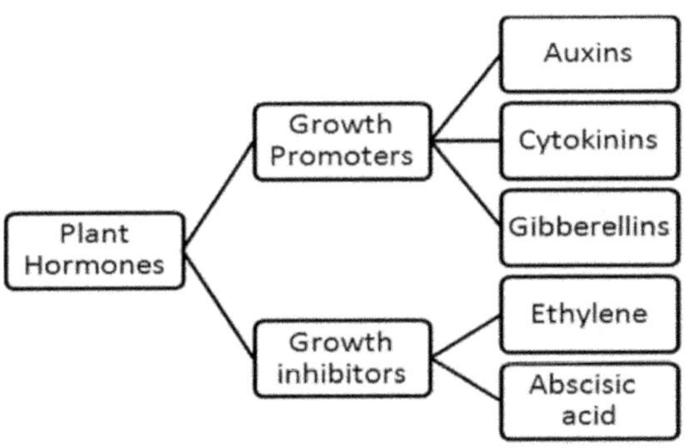

Growth promoters (auxins, cytokinin, gibberellins) and inhibitors (ethylene, abscisic acid)

33. B is correct.

Mosses likely grow in the shaded ground near a small pond.

Mosses are non-vascular plants in the land plant division Bryophyta (mosses, liverworts, and hornworts).

Mosses are a phylum of small soft plants typically 0.4–4 inches (some species grow larger) that commonly grow close together in clumps in damp or shady locations.

Mosses produce spore capsules at certain times.

Mosses differ from vascular plants because they *lack* water-bearing xylem tracheids or vessels.

Mosses do not have flowers or seeds; their simple leaves cover the thin, wiry stems.

Small, non-woody plants (e.g., moss) absorb water and nutrients mainly through their *leaves* and harvest sunlight to create food through photosynthesis.

34. A is correct.

When environmental conditions are unfavorable, is it advantageous for green algae to reproduce sexually.

Sexual reproduction produces offspring with

1) *greater genetic variety* (i.e., recombination during prophase I of meiosis) for survival, and

2) *durable zygotes* help green algae survive unfavorable conditions.

35. D is correct.

Cacti often have large stems that function to produce *food via photosynthesis*.

Cacti survive in hot, dry climates. Most have thick, fleshy stems with waxy skin that store water and keep it from evaporating.

Plants undergo photosynthesis by collecting carbon dioxide through stomata (regulated openings) in leaves and converting it into sugar and oxygen with the input of energy (i.e., photons) from the sun.

Cacti utilize CAM photosynthesis, a process unique to succulents whereby stomata open only at night when the plant is relatively cool, so less moisture is lost through transpiration.

Photosynthesis requires sunlight, so water intake and light-stimulated photosynthesis are separated in time. Cactus leaves are reduced to thin spines to prevent water loss through transpiration. Cacti spines cannot perform photosynthesis carried out by the thick green stem.

Copyright © Sterling Education

36. C is correct.

Phloem is living tissue in vascular plants that transports (i.e., translocation) organic nutrients (e.g., sucrose) synthesized during photosynthesis to vascular plant parts.

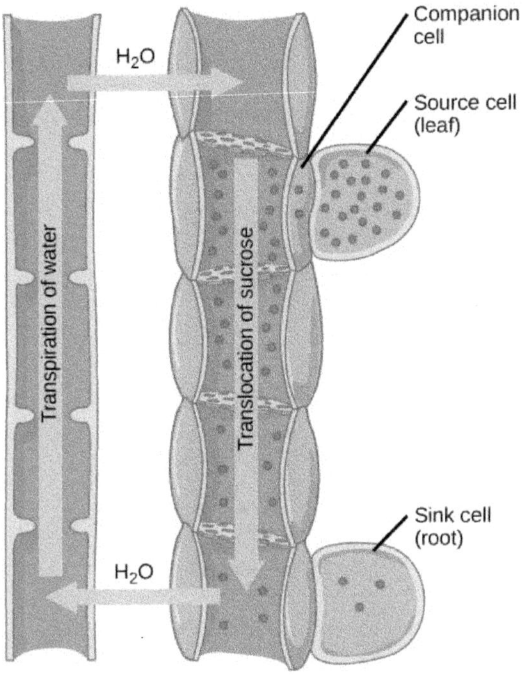

Xylem (on the left) functions for the transpiration of H_2O from roots, with phloem (on right) for the translocation of sugars from leaves during photosynthesis.

Phloem transports (or translocation) *soluble organic material* made during photosynthesis.

Phloem tissue consists of:

conducting cells (i.e., sieve elements),

parenchyma cells (specialized companion and albuminous cells and unspecialized cells), and

supportive cells (i.e., fibers and sclereids).

For example, the phloem is the innermost layer of bark in trees.

A: *lignin* is an alkyl-aromatic polymer in the cell walls of terrestrial plants.

Lignin provides structure and rigidity, is an effective barrier against microbes, and enables water and nutrient transport through plant tissues.

continued...

B: *tracheids* are elongated cells transporting water and mineral salts through the xylem of vascular plants.

Tracheids provide physical support and conduct water and dissolved minerals upwards and are the only such elements in conifers and ferns.

D: *xylem* is in *vascular bundles* in non-woody plants and *non-woody parts* of woody plants.

Xylem transports water and soluble mineral nutrients *from the roots* and replaces water lost during transpiration and photosynthesis.

E: *rhizoid* is a structure in plants and fungi functioning like a root for *support or absorption.*

37. A is correct.

Unlike roots, stems may be involved in *photosynthesis*.

For example, many cacti have large stems that function to produce *food via photosynthesis*.

Photosynthesis

$6CO_2 + 6H_2O \longrightarrow C_6H_{12}O_6 + 6O_2$

carbon dioxide + water ⟶ glucose + oxygen

Photosynthesis reaction:

$$6\ CO_2 + 6\ H_2O \rightarrow C_6H_{12}O_6 + 6\ O_2$$

Six carbon dioxide (CO_2) and six water (H_2O) molecules are converted by light energy (photons) captured by the photopigment chlorophyll into a sugar molecule (e.g., $C_6H_{12}O_6$ as glucose), and six expired oxygen (O_2).

Stems are parts of plants that bear branches, leaves, and fruits.

Roots are underground structures with root hairs.

Roots supply minerals, water, and minerals from the soil and provide *anchorage* for the plant.

38. E is correct.

During primary growth, a stem *increases in length.*

Growth occurs as the stems and roots lengthen.

Apical (shoots and roots) and lateral (cambium) meristems are regions of mitotic cell growth

Some plants, especially woody plants, also increase *thickness* during their life span.

Primary growth is the increase in the length of shoots and roots resulting from cell division in the shoot *apical meristem*.

Copyright © Sterling Education

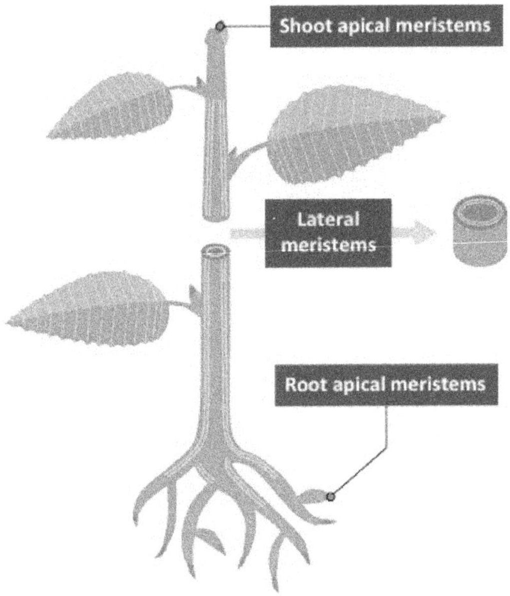

Apical Meristems
Causes **primary** growth *(i.e. lengthening of plant)*
Occurs at tips of shoots and roots
Produces new leaves and flowers

Lateral Meristems
Causes **secondary** growth *(i.e. widening of plant)*
Occurs at the cambium
Produces bark on trees

Meristematic tissues consist of three types, based on location, and contribute to *primary* (taller/longer) and *secondary* (wider) growth.

Intercalary *meristems* (or *basal meristems*) occur only in some monocots, at the bases of leaf blades, and at nodes (the areas where leaves attach to a stem).

This tissue enables the monocot leaf blade to increase in length from the leaf base. For example, grass leaves can elongate even after repeated grazing or mowing.

Apical meristem is the tissue at the tips of stems and roots, enabling length growth.

Lateral meristem facilitates growth in thickness or girth in a maturing plant.

Primary growth is controlled by *root apical* or *shoot apical meristems.*

Secondary growth is controlled by two lateral meristems, *vascular cambium* and *cork cambium.*

Not all plants exhibit secondary growth.

39. C is correct.

Auxins are hormones for cell elongation and are produced in rapidly growing regions near the tip of roots or stems. They coordinate plant development at the cellular level, through organs, and throughout the plant.

Auxin molecules may trigger responses directly through *gene expression* or inhibition.

continued...

Any duplication (copies, uploads, PDFs) is illegal.

Auxin concentration level (along with other local factors) results in cell differentiation and specification. They are essential for cell growth on the cellular level, affecting both cell division and cellular expansion.

A: *cytokinins* are a class of plant hormones promoting cell division (or cytokinesis) in roots and shoots.

They are involved primarily in cell growth and differentiation but also affect apical dominance, axillary bud growth, and leaf senescence.

B: *gibberellins* are plant hormones that stimulate shoot elongation, seed germination, and fruit and flower maturation.

For example, maturing grapes are treated with gibberellins to promote fruit size.

D: *ethylene* is produced from most parts (e.g., leaves, stems, roots, flowers, fruits, tubers, seeds) of higher plants, and production is induced during growth (e.g., germination), fruit ripening, cutting off leaves, and senescence of flowers.

For example, fruit ripening uses *ethylene* for texture changes, softening, color, and other ripening processes.

Ethylene levels are low in under-ripe fruit.

E: *sepals* are part of the flower of *angiosperms* (i.e., flowering plants); morphologically, sepals and petals are modified leaves.

40. B is correct.

Thin tree ring indicates a year of drought (e.g., adverse environmental factor).

Thicker rings indicate a more prolonged or faster growing season when the temperature was high longer and favored growth.

Thinner rings show shorter or slower growing seasons; either the temperature was low, or the tree lacked water.

41. D is correct.

Phototropism is the orientation of an organism (e.g., plant) in response to light.

Phototropism is observed in plants and other organisms, such as fungi.

positive phototropism is toward the light source.

negative phototropism is away from the light source.

Cells farthest from the light contain auxin for phototropism. The plant elongates cells on the furthest side from the light.

continued…

Copyright © Sterling Education

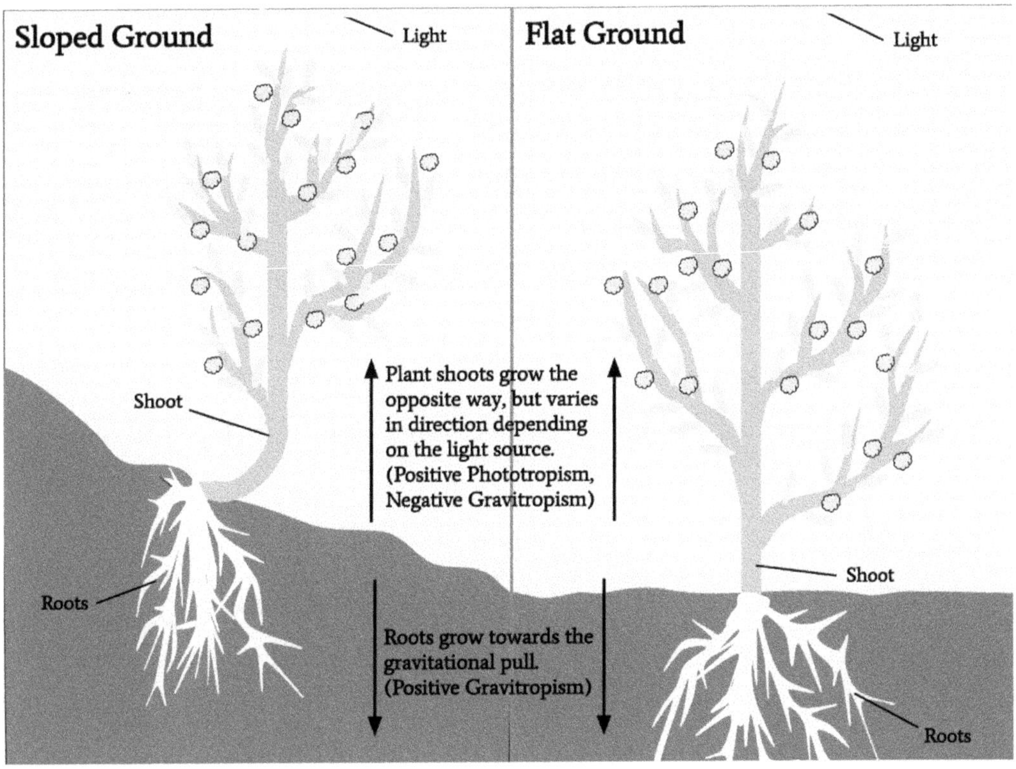

Roots and shoots grow relative to soil pitch for sloped and flat ground. Shoot grows towards the light (most likely up), and roots grow down. Plants on sloped ground (on left) have shoots that bend opposite the gravitational pull, and roots grow down towards the gravitational pull.

Phototropism is a *tropism* (or movements) that responds to external stimuli.

A: *thigmotropism* is a directional growth occurring as a mechanosensory response to a touch stimulus.

Thigmotropism is typical in twining plants and tendrils; also observed in flowering plants and fungi.

B: *rapid movement* encompasses movement in plant structures over a very short period, usually under 1 sec.

For example, the Venus flytrap closes within 100 milliseconds while Utricularia closes in about 0.5 msec.

C: *gravitropism* is a growth response to the environment that directs shoots upward and roots downward, allowing each organ to reach environments adequate for their primary functions.

E: *gibberellins* are plant hormones that regulate developmental processes of stem elongation, germination, dormancy, flowering, flower development, and leaf and fruit senescence.

Any duplication (copies, uploads, PDFs) is illegal.

42. E is correct.

Bryophytes must live in moist areas because they lack vascular tissue, and sperm needs water for the *flagellated sperm* to swim to an egg.

Bryophyte is an embryophyte (i.e., land plant) non-vascular plant without vascular tissue (mosses, liverworts).

Bryophytes are distinct from other land plants because they do *not contain xylem*, the tissue used by vascular plants to transport water.

Instead, bryophytes receive *water and nutrients through their leaves.*

Mosses and liverworts (i.e., Bryophytes) are restricted to moist habitats; deserts have no mosses.

43. A is correct.

Seed formation is a reproductive process in seed plants (*spermatophytes*).

Seed is a structure that includes a plant embryo, endosperm (a food supply), and seed coat (protective covering).

Embryo is a young multicellular organism before it emerges from the seed.

Endosperm is the nutritive tissue of the seed, often a combination of starch, oil, and protein.

Seed coat consists of one or more protective layers encasing the seed.

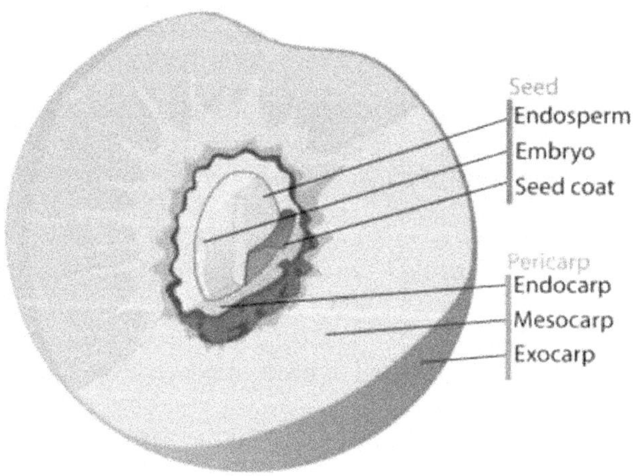

Flowering plant fruit anatomy with seed and pericarp components

B: *gametophyte* is the sexual phase with haploid (1N) male and female organs (or *gametangia*) developing and producing eggs and sperm gametes through mitosis as sexual reproduction.

continued...

 Copyright © Sterling Education

C: *pollen grain* is a single cell containing two male gametes with small haploid (1N) cells encased in a protective coat to prevent desiccation (i.e., drying) and mechanical damage.

Pollen is an immature, endospore male gametophyte derived from male spores (or *microspores*) in seed plants.

Pollen grains are produced by seed plants (i.e., angiosperms and gymnosperms) and are male gametes by meiosis in the anther sac, producing four *microspores* (or pollen grains).

Microspores (or *pollen grains*) mature, the anther opens, and pollen is released.
continued…

D: *spore* is a haploid cell derived from sporangium via meiosis.

Spores are produced by ferns, lycopods, horsetails, mosses, and fungi.

E: *xylem* is in vascular bundles (non-woody plants and non-woody parts of woody plants), secondary xylem laid down by vascular cambium (meristem) in woody plants, and non-bundle arrangements (e.g., ferns).

Xylem transports water, soluble minerals, and nutrients from roots and replaces water from photosynthesis.

44. C is correct.

When positioned horizontally, a plant's response to gravity is that auxins in the lower sides of stems cause cell elongation that bends the stem upright.

Tropism is a biological phenomenon indicating an organism's growth or turning movement (e.g., a plant) in response to an environmental stimulus.

Gravitropism is a movement or growth in response to gravity.

Gravitropism is a growth response directing shoots upward and roots downward, allowing each organ to reach good environments for performing its primary functions.

Auxins are hormones for cell elongation produced in rapidly growing regions near the tip of roots or stems and stimulate stem elongation by promoting cell growth and are a major growth factor in plants.

45. E is correct.

Ethylene is produced from almost all parts (e.g., leaves, stems, roots, flowers, fruits, tubers, seeds) of higher plants, and production is induced during certain stages of growth, such as germination, ripening of fruits, cutting off leaves, and senescence of flowers.

Developmental and environmental factors regulate ethylene production.

continued...

Any duplication (copies, uploads, PDFs) is illegal.

For example, ethylene can be induced by wounding, environmental stresses, and certain chemicals, including auxin and other regulators.

Fruit ripening uses *ethylene* for texture changes, softening, color, and other ripening processes.

For example, the ethylene level in under-ripe fruit is very low.

46. C is correct.

Stomata of leaves are usually open during the day if a plant has enough water.

Stomata are composed of a pair of specialized epidermal cells known as *guard cells.*

They regulate gas exchange (e.g., CO_2) and control water loss by changing the size of the stomatal pore.

Stomata are microscopic cell structures in the epidermis of tree leaves (and needles) involved in exchanging carbon dioxide and water between plants and the atmosphere, which is critical for photosynthesis.

47. B is correct.

Yellow, orange, and red colors of fall leaves result from *anthocyanin* and the *breakdown of chlorophyll.*

Chlorophyll breaks down, the green color disappears, and the yellow to orange colors become visible, giving leaves their fall splendor.

Other chemical changes may occur, forming additional colors by the development of red anthocyanin pigments.

Anthocyanins are colored water-soluble pigments responsible for the colors red, purple, and blue in fruits and vegetables. Berries, currants, grapes, and some tropical fruits contain high anthocyanins.

Carotene is a pigment that absorbs blue and indigo light and provides rich yellows and oranges.

Distinctive colors of mango, fall leaves, yams, yellow butter, and animal fats are due to carotene.

48. A is correct.

O_2 and CO_2 move in and out of a leaf through stomata.

Stomata (singular *stoma*) is a pore in the epidermis of leaves and stems used to control gas exchange.

continued...

 Copyright © Sterling Education

Stoma pore is bordered by a pair of specialized *parenchyma cells* (i.e., *guard cells*) to regulate the size of the pore opening.

Stoma refers to a stomatal complex, the pore, and its accompanying *guard cells.*

49. D is correct.

Growth of plant seedlings is usually *phototropic* and *gravitropic.*

Tropism is a biological phenomenon indicating an organism's growth or turning movement (e.g., a plant) in response to an environmental stimulus.

Phototropism is the orientation of a plant or organism in response to light, either toward the source of light (*positive*) or away (*negative phototropism*).

Phototropism is observed in plants but can occur in organisms such as fungi.

Cells farthest from light contain auxin hormones that react during phototropism.

Gravitropism is a plant growth response directing shoots upward and roots downward, allowing each organ to reach environments adequate for performing its primary functions.

It is a collaborative process of differential growth by a plant or fungi in response to *artificial* or *natural gravity.*

It is a general feature of all higher and many lower plants.

negative gravitropism is the upward growth of plant stems.

positive gravitropism is the downward growth of plant roots.

Primary roots and portions of the root system show positive geotropism by growing towards gravity.

Thigmotropism is a movement or growth in response to touch or contact.

50. B is correct.

Vascular plants are larger than nonvascular plants because *vascular tissue carries water and nutrients* much more efficiently than diffusion. Nutrients and waste products are transported within larger plants over longer distances more efficiently.

The development of vascular tissue allowed plants to grow much taller than nonvascular plants, forming ancient swamp forests. Most of these large vascular seedless plants are extinct, but their smaller relatives remain.

Vascular plants evolved about 420 million years ago (probably) from moss-like bryophyte ancestors, but they had a life cycle dominated by the diploid sporophyte generation.

continued...

Any duplication (copies, uploads, PDFs) is illegal.

Early vascular plants evolved as more plant-like:

- Vascular plants evolved true *roots* made of vascular tissues.

  Compared with rhizoids, roots can absorb more water and minerals from the soil. They securely anchor plants in the ground to grow larger without toppling.

- Vascular plants evolved *stems* made of vascular tissues and *lignin.*

  Lignans stiffen the stems, so plants can grow high above the ground, where they can get more light and air. Because of vascular tissues, stems keep even tall plants supplied with water, so they do not dry out in the air.

- Vascular plants evolved *leaves* to collect sunlight.

  Initially, leaves were tiny and needle-like, which helped reduce water loss. Later, leaves were much larger and broader, so plants could collect more light and undergo photosynthesis to create food.

Early vascular plants had a selective advantage over nonvascular plants with vascular tissues and other adaptations to grow tall and outcompete others for sunlight.

Bryophytes were the photosynthetic pioneers onto land, but early vascular plants were the photosynthetic pioneers into the air.

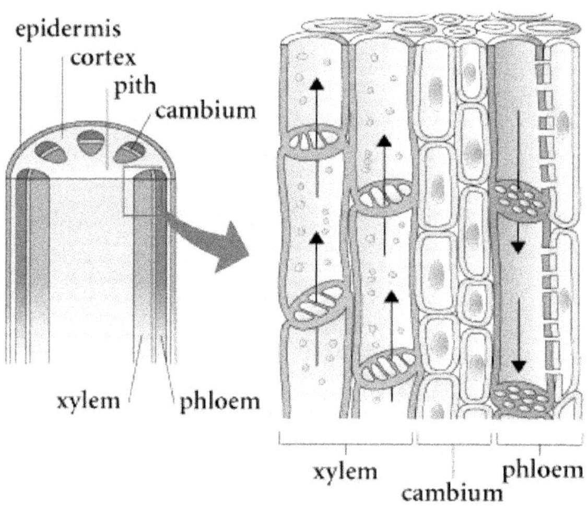

Vascular tissue: xylem (on left) functions for the transpiration of H_2O from roots with phloem (on right) for translocation of sugars from leaves during photosynthesis.

 Copyright © Sterling Education

51. C is correct.

Photosynthesis in plants occurs mainly in *mesophyll tissue*.

Mesophyll tissue (i.e., chlorenchyma or parenchyma) makes up most of the leaf interior between the upper and lower layers of epidermis.

Assimilation (i.e., nutrient absorbing) tissue is the primary location of plant photosynthesis, with the products of photosynthesis, called *assimilates* (e.g., glucose, fructose).

Mesophyll tissue is internal ground tissue between two epidermal cell layers of leaves composed of:

palisade parenchyma (upper layer of elongated chlorenchyma cells with many chloroplasts)

spongy parenchyma (lower layer of spherical or ovoid cells with few chloroplasts and prominent intercellular air spaces).

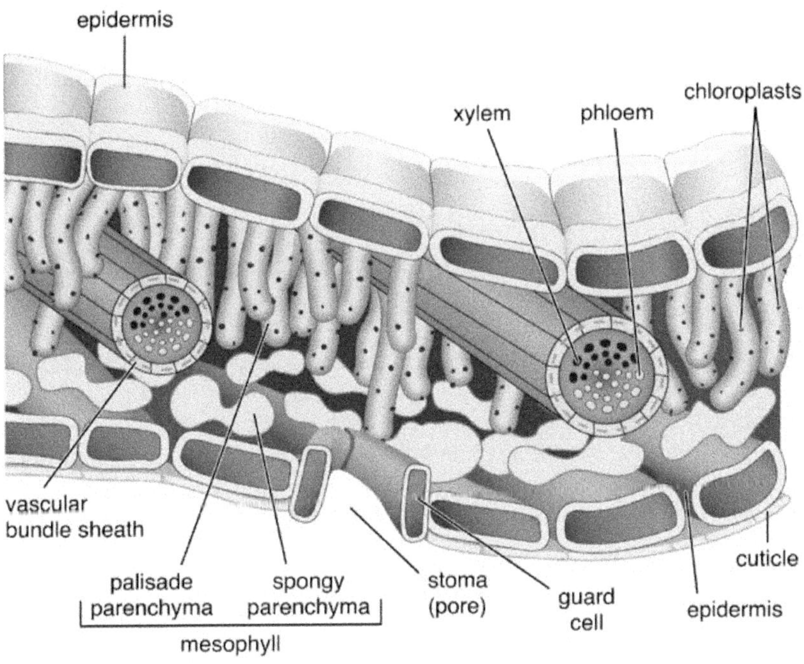

Leaf cross-section with organelles and cellular components

52. A is correct.

Stomata are only on the lower surface of leaves in many plants because water loss would be less on the shaded lower surface than in the direct sun.

Stomata (singular *stoma*) is a pore in the epidermis of leaves and stems for O_2 and CO_2 gas exchange.

continued...

The stomata pore is bordered by a pair of specialized *parenchyma cells* (i.e., guard cells) to regulate the size of the pore opening.

Stoma also refers to a *stomatal complex* (pore accompanying structures).

53. E is correct.

Growth of ivy tendrils wrapping around objects is *thigmotropism,* a growth response to touch or contact.

Tropism is a biological phenomenon of an organism's growth or turning movement (usually a plant) in response to an environmental stimulus.

There are many types of tropisms.

A: *photoperiodism* is not a type of tropism but the physiological reaction of organisms to the length of day or night. It occurs in plants and animals.

B: *chemotropism* is a movement or growth in response to chemicals.

C: *phototropism* is a movement or growth in response to lights or colors of light.

D: *gravitropism* is a movement or growth in response to gravity.

54. B is correct.

Seed-bearing plants differ from other plants because their gametes do *not* require water for fertilization.

*Spermatophyte*s (i.e., seed plants) include *gymnosperms* (conifers) and *angiosperms* (flowering plants).

Seeds are produced by flowering *plants* (angiosperms) and cone *plants (gymnosperms).*

Spores are produced by ferns, mosses, and green algae and require water for fertilization.

Gymnosperms are a group of seed-producing plants (e.g., conifers, cycads) with seeds developing on the surface of scales or leaves, often modified to form cones or at the end of short stalks (e.g., Ginkgo).

Gymnosperm seeds are not enclosed within an ovary, unlike *angiosperms* (flowering plants).

Both male (*pollen*) gametes are involved in fertilization, forming a *zygote* (or fertilized egg) and an endosperm; double fertilization by two gametes is unique to flowering plants.

Copyright © Sterling Education

55. D is correct.

Xylem tissue in vascular plants is vital because it conducts water over long distances.

Xylem is in vascular bundles (non-woody plants and non-woody parts of woody plants), secondary xylem laid down by vascular cambium (meristem) in woody plants, and non-bundle arrangements (e.g., ferns).

Xylem transports water, soluble minerals, and nutrients from roots and replaces water lost during photosynthesis.

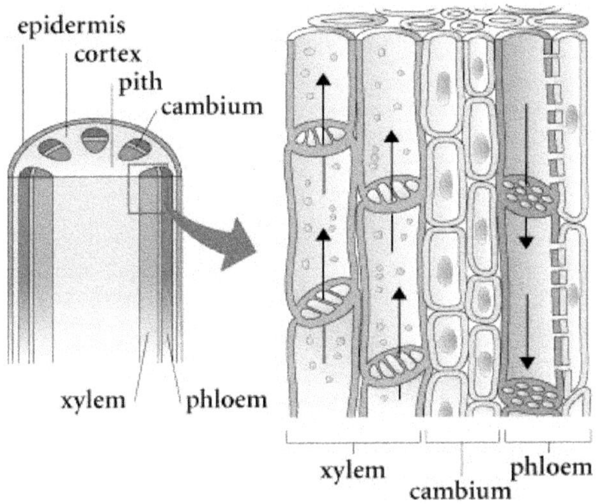

Xylem functions for transpiration of H_2O from roots

56. B is correct.

Corn, broccoli, and cabbage were developed by selective breeding.

Selective breeding is when humans use animal or plant breeding to selectively develop phenotypic traits by choosing representative male and female gametes to reproduce sexually and produce offspring.

Many fruits and vegetables have been bred through artificial selection.

For example, by selective breeding, broccoli, cauliflower, and cabbage were derived from wild mustard plants.

A: *auxins* are plant hormones for cell elongation produced in rapidly growing regions near the tip of roots or stems. They stimulate stem elongation by promoting cell growth and are a major growth factor in plants.

C: *plant propagation* is reproducing plants from a single parent plant.

continued...

Any duplication (copies, uploads, PDFs) is illegal.

Asexual propagation includes cuttings, layering, division, budding, and grafting.

cuttings involve rooting a severed piece of the parent plant;

layering involves rooting a part of the parent and then severing it; and

budding and *grafting* join two plant parts from different varieties.

E: *pollination* transfers pollen grains from the *male anther* of a flower to the *female stigma.*

57. A is correct.

In the figure, the water pressure is high in *guard cells*, specialized cells in the lower leaf epidermis of plants.

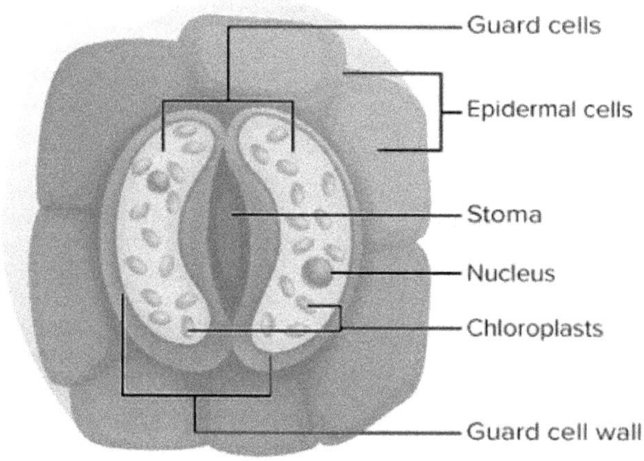

Leaf underside cross-section with an open stoma and support structures

Pairs of guard cells surround stomata pores and regulate CO_2 intake from the atmosphere for photosynthesis.

Guard cells also regulate how much water plants lose to the atmosphere.

Opening and closing the stomata pores uses changes in the turgor (i.e., osmotic) pressure of two guard cells.

When water flows into the guard cells, they swell, and the curved surface causes the stomata to open.

When the guard cells lose water, they shrink and become flaccid and straight, thus closing the stomata.

 Copyright © Sterling Education

58. E is correct.

Xylem cells move water by *capillary pressure.*

Due to transpiration pull, when transpiration is high, xylem sap is under tension rather than pressure.

Capillary action is the movement of water within spaces of a porous material due to *adhesion, cohesion,* and *surface tension,* which is an elemental force for moving water and dissolved molecules.

> *Adhesion* is the attraction of molecules among different kinds of molecules and is strong for water, especially including positive or negative charges molecules.
>
> *Cohesion* (water sticking to each other) causes more water molecules to fill the gap in the xylem as the topmost water is pulled toward the stomata.
>
> *Surface tension* created by transpiration *pulls* water upward in the plant xylem.

A: *mesophyll cells* (i.e., chlorenchyma or parenchyma) are the internal ground tissue comprising most of the leaf interior between the upper and lower layers of the epidermis.

B: *albuminous cells* are specialized parenchyma cells and major component of ground tissue in vascular plants.

Albuminous cells and albumin protein are associated with phloem sieve tube cells in gymnosperm wood.

C: *phloem cells* are living tissue in vascular plants that transports (i.e., translocation) organic nutrients (e.g., sucrose) synthesized during photosynthesis to vascular plant parts.

D: *guard cells* are specialized cells in the lower leaf epidermis of plants.

Pairs of guard cells surround stomata pores and regulate CO_2 intake from the atmosphere for photosynthesis.

Stomata also regulates how much water plants lose to the atmosphere.

59. C is correct.

Pollen grains are a haploid (1N) gymnosperm reproductive structure.

Spermatophytes (i.e., seed plants) include *gymnosperms* and *angiosperms* as two main groups.

Seeds are produced by flowering *plants* (*angiosperms*) and cone *plants (gymnosperms)*, while spores are produced by ferns, mosses, liverworts, and green algae and require water for fertilization.

continued...

Gymnosperms are seed-producing plants (e.g., conifers, cycads, Ginkgo) with the following characteristics:

naked seeds,

separate female and male gametes,

pollination by wind, and

tracheids to transport water and solutes in the vascular system.

Gymnosperm seeds develop on the surface of scales or leaves, often modified to form cones or at the end of short stalks, as in Ginkgo.

Gymnosperm seeds are not enclosed within an ovary, unlike flowering plants (angiosperms).

Gymnosperms (or *naked seeds*) have seeds *exposed* at stalk ends (or *whorls*) or on *cone scales*.

Gymnosperm with ovule, seed, pollen chamber, and aborted spores

Pollen grains mainly represent the (1N) male part of the reproductive process in plants and trees. After forming four *haploid microspores*, they fuse to form *pollen grain,* and the pollen grain walls develop. Therefore, a pollen grain is a haploid (1N) structure.

A: *seeds are produced by* flowering *plants* (*angiosperms*) and cone *plants (gymnosperms).*

Seed is the (2N) embryonic stage of the plant life cycle.

Most seeds consist of three parts:

embryo, *endosperm*, and *seed coat*

Endosperm is the nutritive tissue of the seed, often a combination of starch, oil, and protein.

B: *cones* are the reproductive structure in conifers with modified leaves densely arranged on a central stalk.

Female (or *seed cones*) bear ovules on the surface of their scales, while *male cones* produce pollen.

continued...

 Copyright © Sterling Education

D: *seed cone* is the female fruit of conifers and terminal, cone-like, usually fleshy, and rarely dry.

Seed cones of conifers are compact branching systems composed of two separate structures: a bract (or modified leaf) and an ovuliferous scale (or modified seed-bearing shoot).

Seed-bearing cones are female, while the *pollen-filled cones are male.*

Both sexes of cones grow on the same tree, but the male cones typically grow on lower branches relying on the wind to blow pollen up to the female cones.

E: *embryo* is a tiny (2N) plant with a root, stem, and one or more leaves.

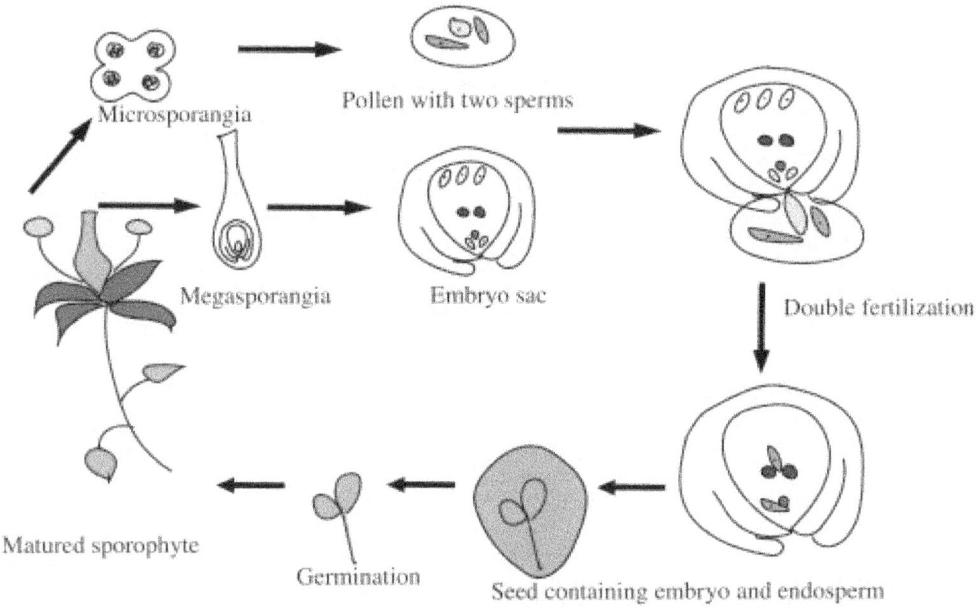

Sporophytes form megasporangia and four microsporangia. The embryo sac undergoes double fertilization to produce a seed containing embryo and endosperm

60. D is correct.

People use plants to provide food, raw materials, and medicines.

Medicinal plants (or *medicinal herbs*) have been used in traditional medicine practices since prehistoric times.

Medicinal plants and herbs (e.g., turmeric, ginger, basil leaves, mint, and cinnamon) are common in culinary dishes and offer health benefits (e.g., cold and flu remedies, relieve stress, improve digestion, and improve immune responses).

Plants synthesize chemical compounds for defense against insects, fungi, diseases, and herbivorous mammals.

Any duplication (copies, uploads, PDFs) is illegal.

61. A is correct.

Embryo develops in the ovule.

Carpel (or pistil) are female structures, while *stamen* are male structures.

Pistil (or *carpel*) is the female organ of a flower, with the stigma, style, and ovary.

Stamen is the male reproductive organ of flowering plants, consisting of the *anther* (site of pollen development) and often a stalk-like *filament* (transmits water and nutrients to the anther and positions it for pollen dispersal).

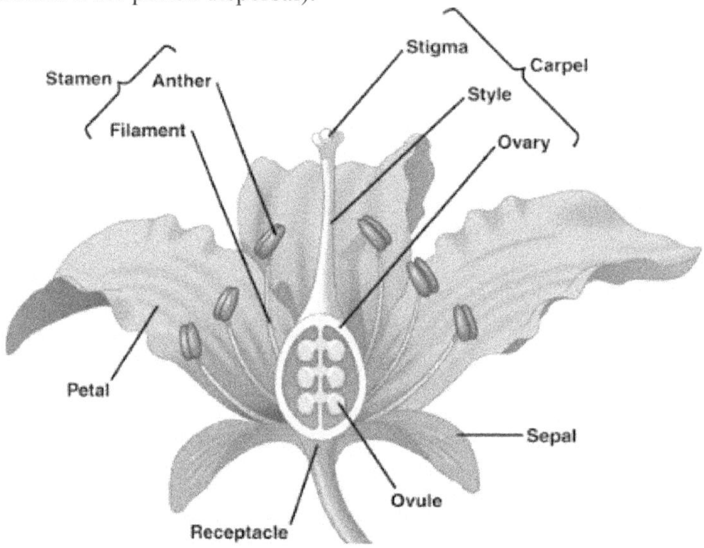

Carpel (or pistil) are female structures, while stamen are male structures

62. B is correct.

Pressure-flow hypothesis (or mass-flow hypothesis) is the best-supported theory to explain food movement through the phloem. German plant physiologist Ernst Munch proposed it in 1930.

Pressure-flow hypothesis proposes that water containing food molecules flows under pressure through the phloem. The pressure is created by the difference in water concentration of the solution in the phloem and the relatively pure water in the nearby xylem ducts.

A: *root pressure* is the transverse *osmotic pressure* within cells of a root system that causes sap to rise through a plant stem to the leaves.

Osmotic pressure is the minimum pressure for a solution to stop the flow of solvent molecules through a semipermeable membrane (osmosis). It is a colligative property and depends on the concentration of solute particles in the solution.

continued...

 Copyright © Sterling Education

C: *transpiration pull* is a biological process in which the force of pulling is produced inside the xylem tissue. This force helps in the upward movement of water into the xylem vessels. In this process, water loss through leaves is observed as vapors.

D: *diffusion gradient* is a difference in the diffusion rates (net movement of a substance from higher to lower concentration) of molecules through a medium or substrate.

E: *capillary action* is the movement of water within the spaces of a porous material due to the forces of *adhesion, cohesion*, and *surface tension* are vital forces for moving water and dissolved molecules.

63. A is correct.

According to the pressure-flow hypothesis, the movement of H_2O into a nutrient-rich region of the phloem *increases* the pressure.

Pressure-flow hypothesis (or *mass-flow hypothesis*) explains the force for food movement through the phloem.

A high concentration of organic substances (e.g., sugars from photosynthesis) inside phloem cells at a source (e.g., a leaf) creates a diffusion gradient (osmotic gradient), drawing water into cells.

Sap originates in xylem cells of sapwood and is mostly water with minerals, sugar, nutrients, and hormones. *Sapwood* is the soft outer layer of recently formed wood between the heartwood and bark, with vascular tissue.

Phloem sap moves from sugar sources (leaves) to sugar sinks via turgor pressure (i.e., hydrostatic pressure).

A sugar source is any part of the plant producing or releasing sugar.

Tree sap can be in sapwood, which creates carbon dioxide. The sap is forced out in openings or wounds if the carbon dioxide creates pressure inside the tree.

During the plant's growth period (usually during the spring), storage organs (e.g., roots) are sugar sources, while many growing areas are sugar sinks.

Water moves sugar from roots to twigs when ambient temperatures rise, supplying the energy to grow new shoots and leaves. Pruning results in bleeding as sap oozes from the wound.

Movement in the phloem is bidirectional, while in xylem cells, it is unidirectional (upward).

64. C is correct.

Angiosperms (or *flowering plants*) enclose seeds in *ovaries* during development.

S*permatophytes* (i.e., seed plants) include *gymnosperms* and *angiosperms* as two main groups.

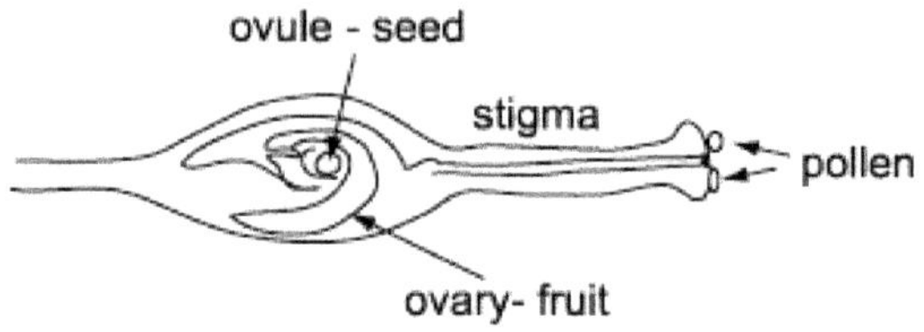

Angiosperm with ovule (seed), ovule (fruit), stigma, and pollen grains

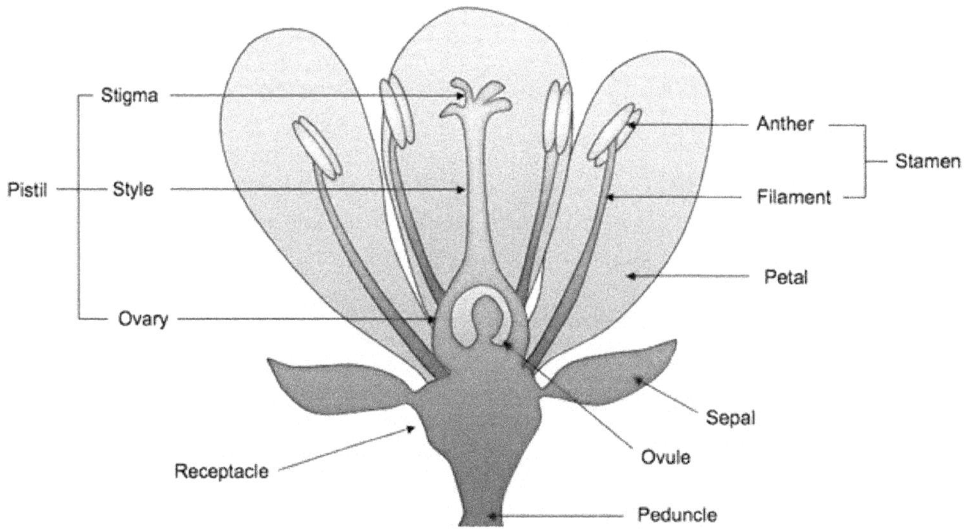

Carpel (or pistil) are female structures, while stamen are male structures

A: *petals* are sterile floral parts as visually conspicuous elements to attract specific pollinators to the flower.

B: *cones* are reproductive structures in conifers with modified leaves densely arranged on a central stalk.

Female (or *seed cones*) bear ovules on the surface of their scales, while male cones produce pollen.

Seed-bearing cones are female, while the pollen-filled cones are male.

Both sexes of cones grow on the same tree, but male cones typically grow on lower branches relying on wind to blow pollen up to the female cones.

continued...

 Copyright © Sterling Education

D: *pollen grains* are produced by seed plants (i.e., angiosperms and gymnosperms) and are male gametes by meiosis in the anther sac, producing four microspores (or pollen grains).

Pollen grains carry male gametes with small haploid (1N) cells encased in a protective coat to prevent desiccation (i.e., drying) and mechanical damage.

E: *stomata* are composed of a pair of specialized epidermal cells known as *guard cells* that regulate gas exchange (e.g., CO_2) and control water loss by changing the size of the stomatal pore.

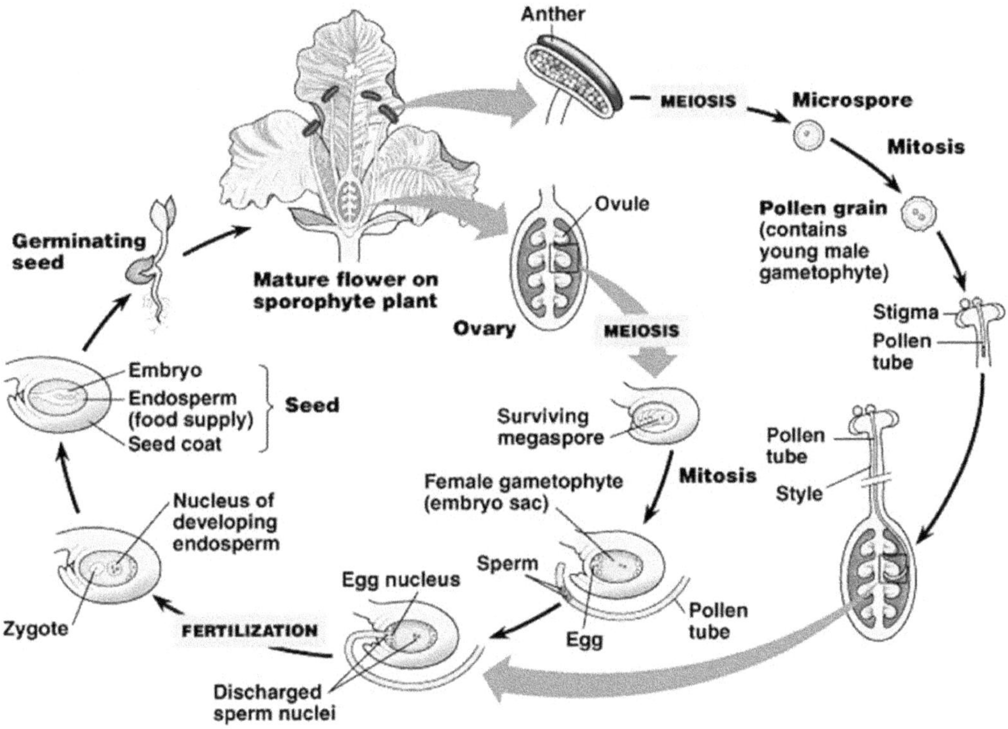

Angiosperm life cycle with microspore and megaspore undergoing fertilization

65. E is correct.

Ovaries develop into fruit in angiosperms. As the seed develops in an ovary, the walls of the ovary thicken and form the fruit.

A fertilized and fully grown, ripened ovary is a fruit.

continued...

Any duplication (copies, uploads, PDFs) is illegal.

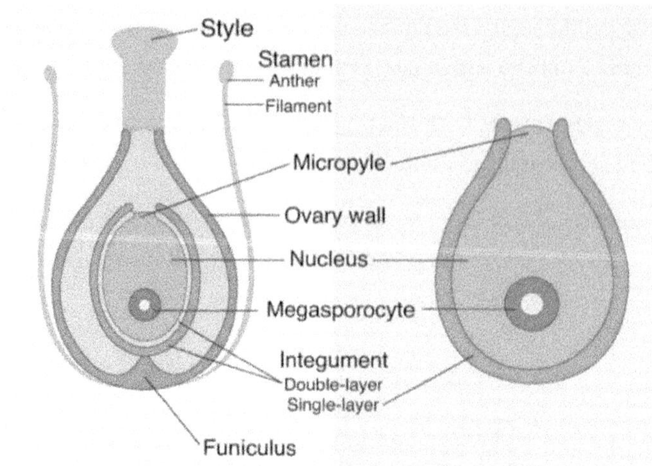

Angiosperm flowering plants (on left) and gymnosperm (on right) anatomy

66. D is correct.

Pollen grain landing on a stigma of a flower produces a *pollen tube*.

Pollination occurs when pollen lands on a compatible *pistil* or *female cone*. It germinates and produces a *pollen tube* that transfers the sperm to the *ovule* (contains female gametophyte).

Pollen grains are microscopic (require magnification to visualize) fine to coarse powder containing microgametophytes that produce male gametes (i.e., sperm cells).

Pollen grains have a hard coat that protects sperm cells moving from the:

> *stamens* (i.e., pollen-producing reproductive organ) to the *pistil* (i.e., a tube-like portion between stigma and ovary) of flowering plants, or
>
> *male cone* (i.e., organ with reproductive structure) to the *female cone* of coniferous plants.

Pollen is essential in forensics, archeology, paleoecology, and paleontology (i.e., the study of prehistoric life).

67. D is correct.

Annuals are flowering plants that grow, produce seeds, and die in one year.

A: *biennials* require two years to grow from seed to fruition and death.

B: *perennials* regrow every spring and generally have a shorter blooming period than annuals.

C: *dicots* (or *dicotyledons*) are one of two flowering plant (*angiosperms*) groups.

continued...

Copyright © Sterling Education

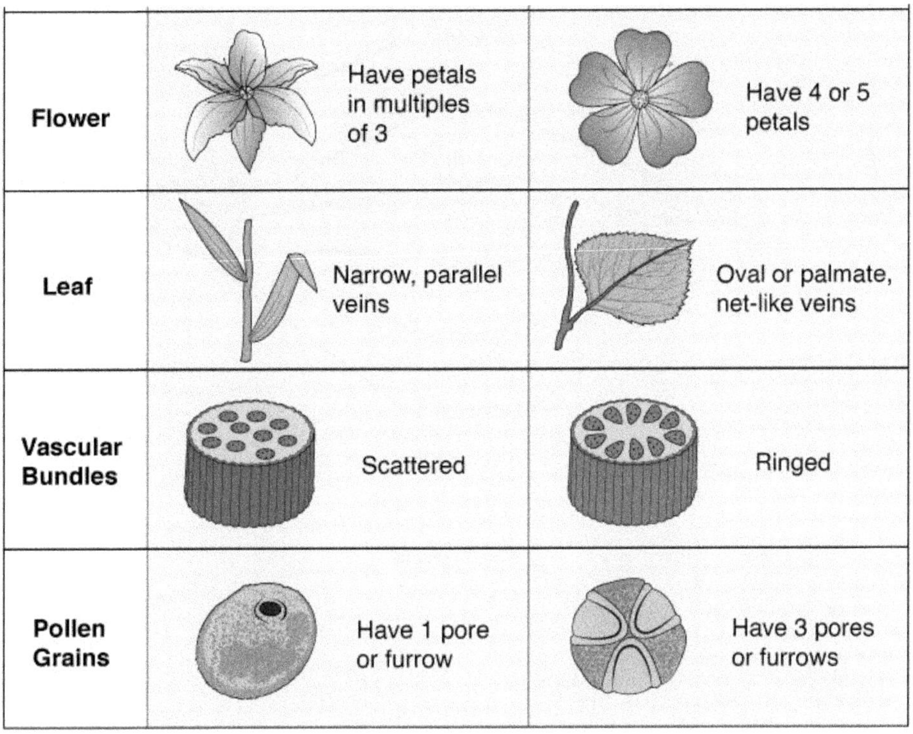

Monocots (on the left), compared to dicots (on the right), have distinctions in flowers, leaves, vascular bundles, pollen grains, cotyledons, roots, stems, and roots

Seed with two *embryonic leaves* (or *cotyledons*) is a characteristic of the dicots.

Cotyledon is part of the embryo within the seed and the embryonic leaf in seed-bearing plants, appearing first from a germinating seed.

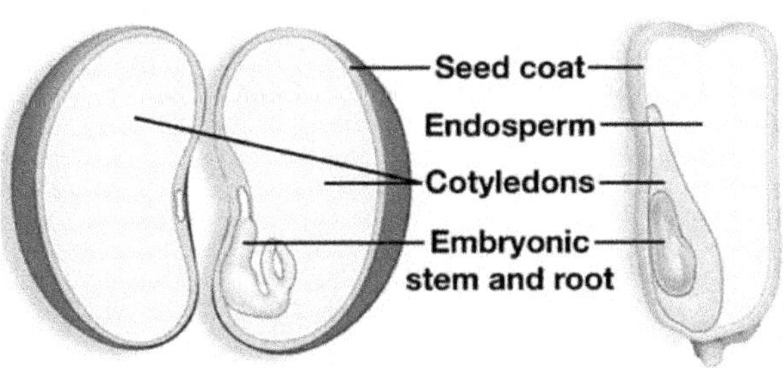

Bean seed (left) and corn seed (right) cotyledons and embryonic stem and root comparison

Cotyledons number is a characteristic used by botanists to classify flowering plants.

E: *monocots* are grass and grass-like flowering plants (angiosperms), the seeds of which typically contain only one embryonic leaf or cotyledon.

continued...

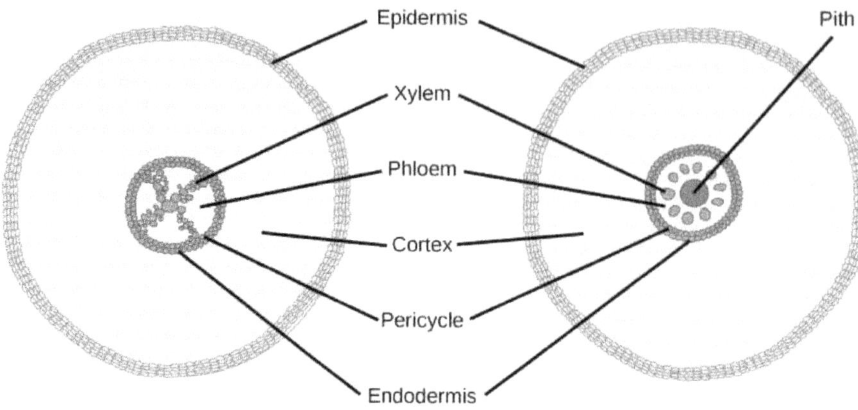

Dicot root (left) and monocot root (right) with scattered or ringed vascular bundles

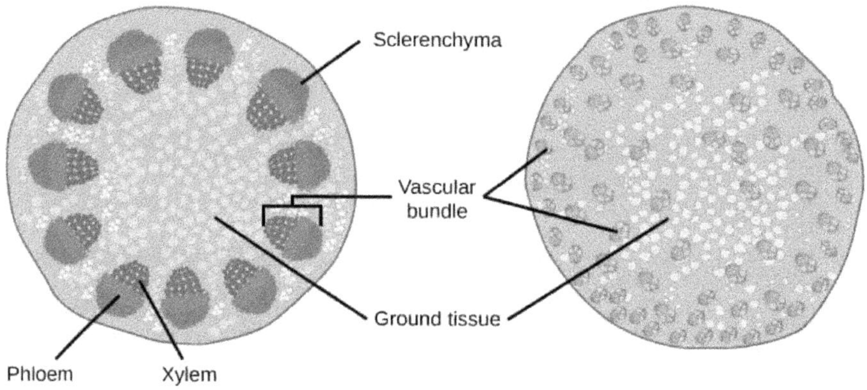

Dicot stem (left) compared to monocot stem (right) with scattered or ringed vascular bundles

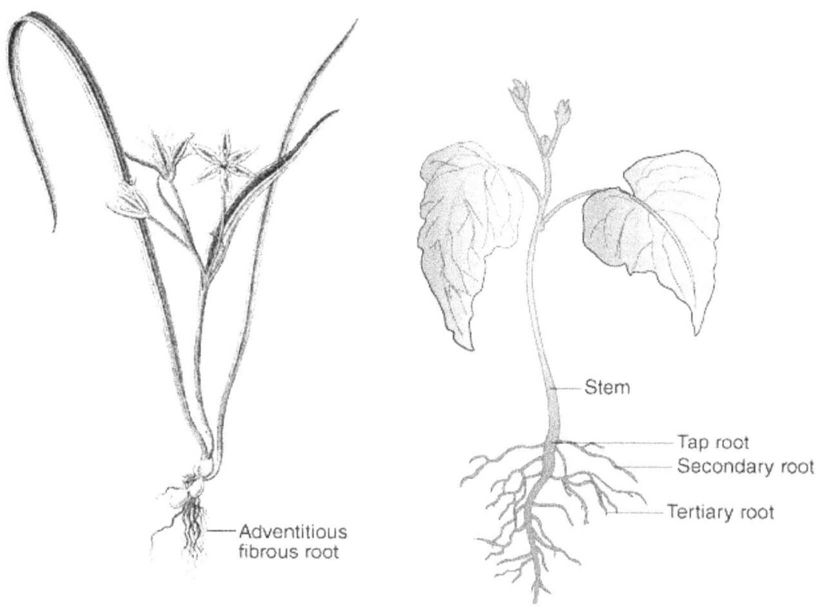

Monocot root (left) and dicot root (right)

Copyright © Sterling Education

68. B is correct.

Ground tissue is in roots, stems, leaves, and all tissues that are neither dermal nor vascular.

Ground tissue by three types of cell walls.

Parenchyma cells have *thin primary walls* and usually remain alive after they mature, which forms the *filler tissue* in the soft parts of plants.

Collenchyma cells have *thin primary walls* with some areas of *secondary thickening* and provide extra *structural support* (i.e., regions of new growth).

Sclerenchyma cells have *thick secondary walls* and often die when mature, providing the primary structural support to a plant.

69. E is correct.

Seed plants are held to the ground by *roots.*

A: *phloem* is living tissue in vascular plants that transports (i.e., translocation) organic nutrients (e.g., sucrose) synthesized during photosynthesis to vascular plant parts.

B: *epidermis* (or d*ermal tissue*) is the outer covering of a plant (roots, stems, leaves, flowers, fruits, and seeds); usually, one cell layer thick, and its cells lack chloroplasts.

C: *stems* are parts of plants that bear branches, leaves, and fruits.

D: *xylem* transports water, soluble minerals, and nutrients from roots and replaces water from photosynthesis.

Xylem is in vascular bundles (non-woody plants and non-woody parts of woody plants), secondary xylem laid down by vascular cambium (meristem) in woody plants, and non-bundle arrangements (e.g., ferns).

Notes for active learning

 Copyright © Sterling Education

Detailed Explanations: Photosynthesis

Answer Key

1: E	11: C	21: A	31: C	41: A	51: A
2: D	12: C	22: B	32: A	42: B	52: B
3: A	13: E	23: E	33: E	43: B	53: D
4: B	14: B	24: C	34: C	44: C	54: C
5: C	15: A	25: B	35: A	45: E	55: B
6: D	16: A	26: E	36: D	46: A	56: E
7: A	17: B	27: D	37: E	47: D	57: C
8: B	18: E	28: D	38: B	48: A	
9: A	19: D	29: C	39: A	49: E	
10: E	20: A	30: A	40: B	50: C	

1. E is correct.

C_4 plants use C_4 carbon fixation.

Three carbon fixation mechanisms exist C_3, C_4 and CAM (*crassulacean acid metabolism*).

C_4 is named the 4-carbon molecule in carbon fixation compared to a 3-carbon molecule product in C_3 plants.

C_4 plants have a competitive advantage over plants that use the C_3 carbon fixation pathway under drought, hot temperatures, and nitrogen or CO_2 limitation conditions.

8,100 plant species use C_4 carbon fixation, representing about 3% of terrestrial plants.

C_4 and CAM overcome the tendency of the RuBisCO to fix O_2 rather than CO_2 in photorespiration. They use a more efficient enzyme to fix CO_2 in mesophyll cells and shuttle this fixed carbon to bundle-sheath cells.

In bundle-sheath cells, RuBisCO is isolated from atmospheric oxygen and saturated with CO_2 released by decarboxylation of malate or oxaloacetate.

C_4 fixation is hypothesized to have evolved later.

Because these additional steps require more energy from ATP, C_4 plants can efficiently fix carbon in specific conditions, with the C_3 pathway being efficient in other conditions.

Any duplication (copies, uploads, PDFs) is illegal.

2. D is correct.

CAM (*crassulacean acid metabolism*) photosynthesis is a carbon fixation pathway that evolved in some plants to adapt to arid conditions.

In a plant using CAM, the stomata in the leaves remain shut during the day to reduce evapotranspiration but open at night to collect carbon dioxide (CO_2).

CO_2 is stored as the four-carbon acid malate and used for photosynthesis during the day.

The pre-collected CO_2 is concentrated around the RuBisCO enzyme,
increasing photosynthetic efficiency.

3. A is correct.

Autotrophs (i.e., *producers*) synthesize complex organic compounds (i.e., carbohydrates, fats, proteins) from simple substances in their environment, generally using energy from light (i.e., photosynthesis) or inorganic chemical reactions (i.e., chemosynthesis).

Autotrophs are the producers in a food chain (plants on land and algae in water), in contrast to heterotrophs that consume autotrophs.

Autotrophs can make their food and do not need living energy or an organic carbon source.

Autotrophs can reduce carbon dioxide to make organic compounds.

Most autotrophs use water as the reducing agent, but some can use other hydrogen compounds.

4. B is correct.

Adenosine triphosphate (ATP) is a *nucleotide* composed of adenosine (A), ribose, and three phosphate groups.

ATP has three phosphate groups and is produced by enzymes (e.g., ATP synthase) from adenosine diphosphate (ADP) or adenosine monophosphate (AMP) and phosphate groups.

Mitochondria are the site of most ATP molecules produced during aerobic respiration.

 Copyright © Sterling Education

5. C is correct.

Photosynthesis reaction:

$$6\ CO_2 + 6\ H_2O \rightarrow C_6H_{12}O_6 + 6\ O_2$$

Reactants are six carbon dioxide (CO_2) and six water (H_2O) molecules, converted by light energy captured by chlorophyll pigments into one sugar ($C_6H_{12}O_6$) and six oxygen (O_2).

Chloroplasts are the site of photosynthesis and are only in algae and plant cells.

Chloroplasts contain DNA and ribosomes and may have similarly evolved via endosymbiosis.

6. D is correct.

ATP is a nucleoside triphosphate with a nitrogenous base (adenine), ribose sugar, and triphosphate.

The complete structure represents ATP.

Structure A represents *adenine* (nitrogenous base).

Structure B represents *ribose*.

Structure C represents *diphosphate*.

Structure D represents *phosphate*

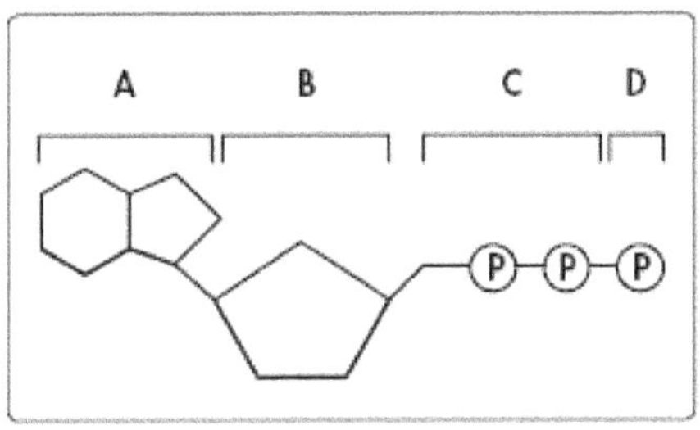

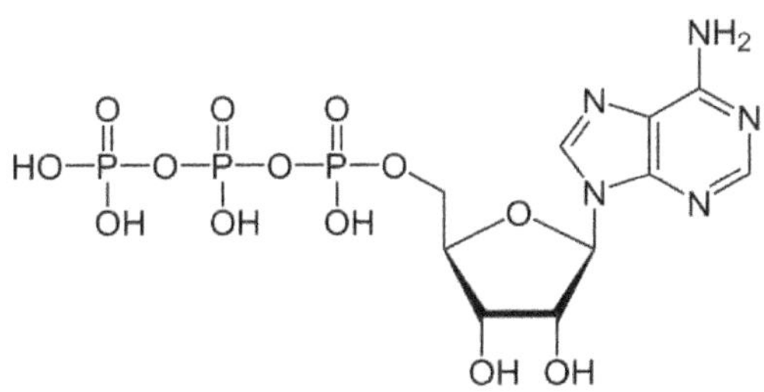

Adenosine triphosphate

7. A is correct.

Stroma is the colorless fluid surrounding the thylakoids within the chloroplast.

Within the stroma are *grana* (i.e., stacks of thylakoids) – the sub-organelles, where photosynthesis begins before the chemical changes are completed in the stroma.

8. B is correct.

During photosynthesis, autotrophs produce carbohydrates (e.g., glucose).

Photosynthesis:

$$6\ CO_2 + 6\ H_2O \rightarrow C_6H_{12}O_6 + 6\ O_2$$

Reactants are six carbon dioxide (CO_2) and six water (H_2O) molecules, converted by light energy captured by chlorophyll pigments into one sugar ($C_6H_{12}O_6$) and six oxygen (O_2) molecules.

9. A is correct.

Stomata are composed of a pair of specialized epidermal cells known as *guard cells*.

Stomata regulate gas exchange (e.g., CO_2) and control water loss by changing the size of the stomatal pore.

10. E is correct.

Plants use photosynthesis to make food.

During photosynthesis, plants absorb photons (light energy) with their leaves.

Plants use energy from the sun to convert water (H_2O) and carbon dioxide (CO_2) into glucose ($C_6H_{12}O_6$), six-carbon chain sugar.

Plants use glucose ($C_6H_{12}O_6$) for cellular energy to make biomolecules (e.g., cellulose and starch in plants; glycogen in animals).

Cellulose builds cell walls, while starch is stored in seeds and plant parts as a food source.

Chlorophylls *a* and *b* are photosynthetic pigments produced in chloroplasts of photosynthetic tissues of leaves.

Chlorophyll molecules are water-repelling because of the long phytol tail in the molecule.

Structures below...

 Copyright © Sterling Education

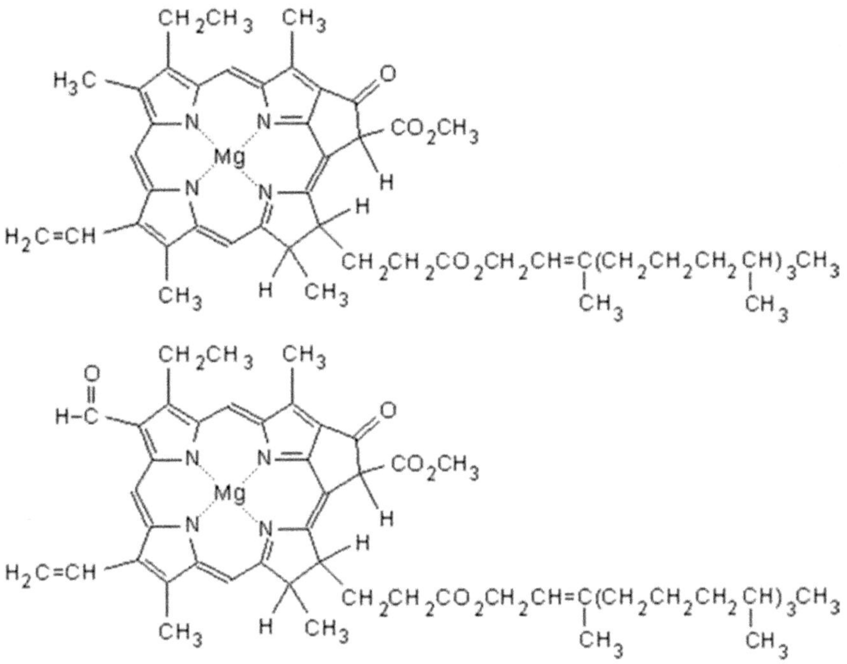

Photopigments chlorophyll a (top) and chlorophyll b (bottom) absorb energy during photosynthesis

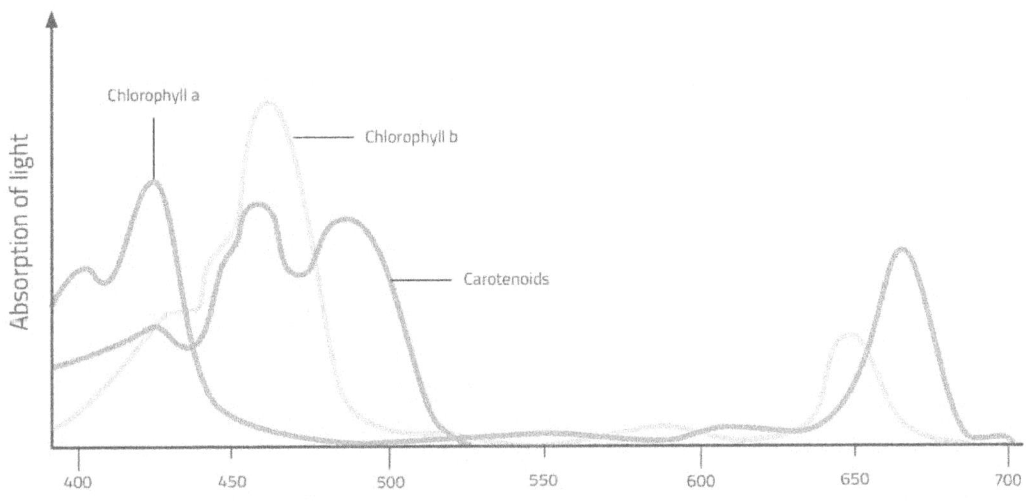

Photopigments absorb specific wavelengths in a visible light spectrum from 380 to 700 nm

11. C is correct.

Roots are used by plants to obtain most of their water.

12. C is correct.

Thylakoids are labeled in the figure.

13. E is correct.

Thylakoids are membrane-bound compartments inside chloroplasts and cyanobacteria.

Thylakoids are the site of photosynthesis.

Thylakoids consist of a thylakoid membrane surrounding a thylakoid lumen.

Thylakoids stack as disks (i.e., grana).

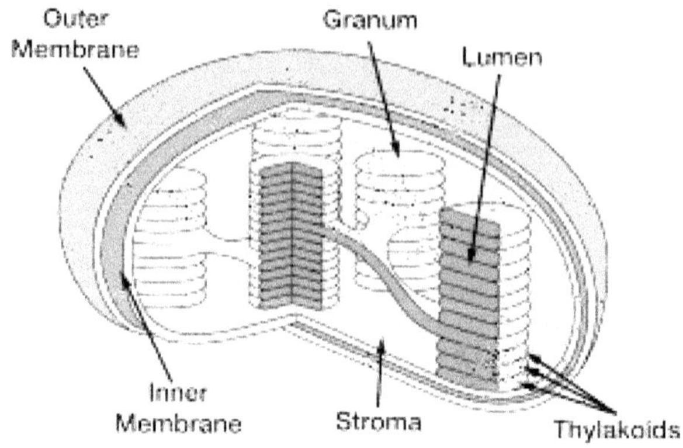

Chloroplast with internal structures

14. B is correct.

Chlorophyll is within the thylakoid membrane.

15. A is correct.

Granum (pl. grana) is a stack of thylakoid discs. Chloroplasts can have from 10 to 100 grana.

Grana are connected by stroma thylakoids (or *lamellae*).

Grana thylakoids and stroma thylakoids are distinguished by their protein composition.

16. A is correct.

Electron carrier molecule in the figure is $NADP^+$.

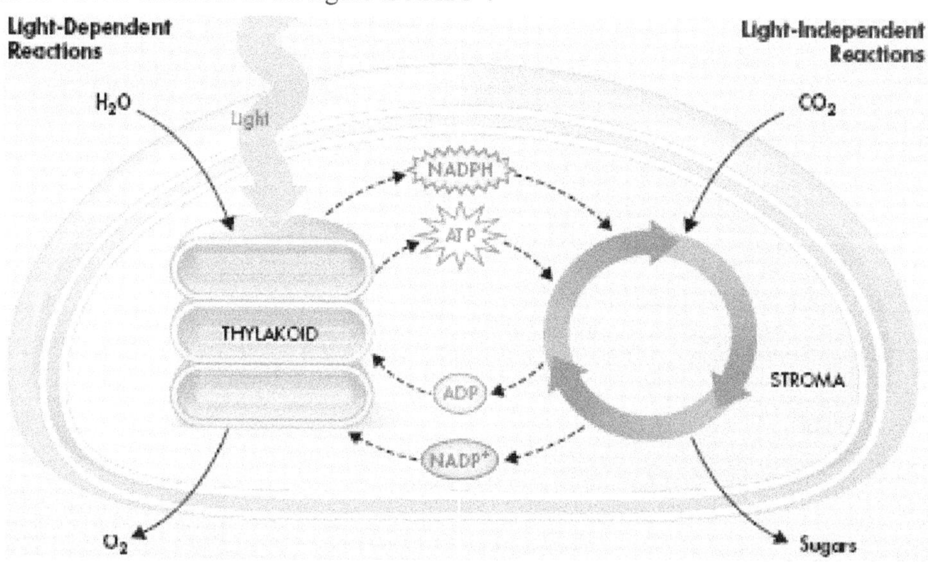

Chloroplast with light-dependent (ATP and NADPH) and light-independent reactions (sugars)

Copyright © Sterling Education

17. B is correct.

Low ATP production is likely if a shade-tolerant plant has minimal water intake.

18. E is correct.

High-energy electrons are highly reactive and are transported in the chloroplast by *electron carriers* ($NADP^+$).

19. D is correct.

Photosynthesis reaction:

$$6\ CO_2 + 6\ H_2O \rightarrow C_6H_{12}O_6 + 6\ O_2$$

Reactants are six carbon dioxide (CO_2) and six water (H_2O) molecules, converted by light energy captured by chlorophyll pigments into one sugar ($C_6H_{12}O_6$) and six oxygen (O_2) molecules.

20. A is correct.

Rate of photosynthesis *increases* and then levels off as the intensity of light increases.

21. A is correct.

In the redox reactions of photosynthesis, *electron transfer* is from $H_2O \rightarrow CO_2$

22. B is correct.

Steps in the light-dependent reaction include ATP synthase allowing H^+ ions to pass through the thylakoid membrane, high-energy electrons through the electron transport chain, and pigments in photosystem II absorbing photons of light.

23. E is correct.

Oxygen is the product of the light reaction of photosynthesis.

24. C is correct.

Photosystems I and II are in the *thylakoid membrane.*

Any duplication (copies, uploads, PDFs) is illegal.

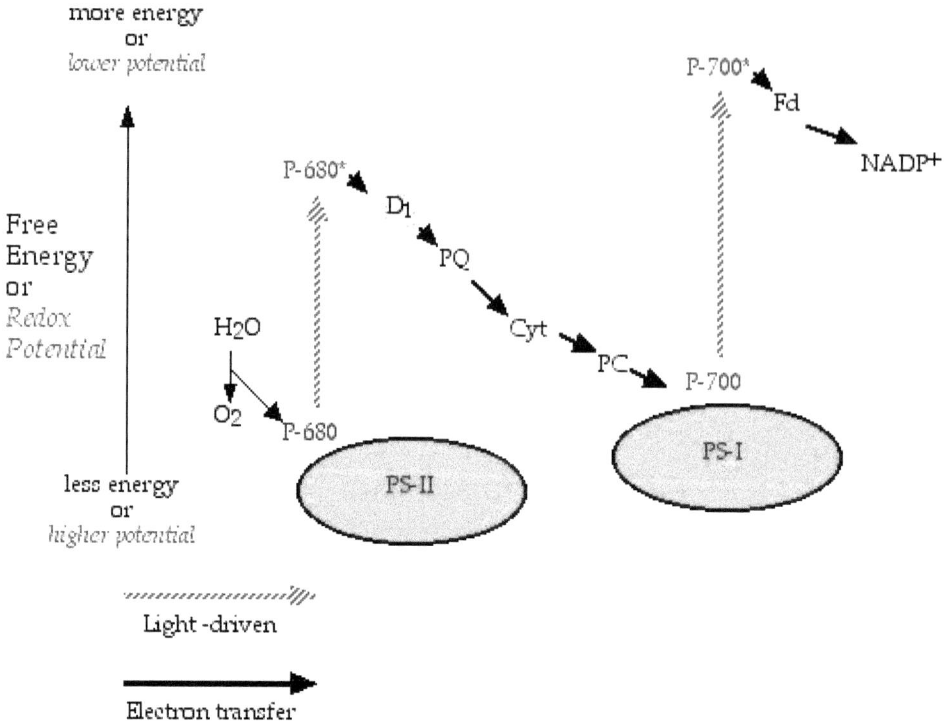

Photosystem II and I in the thylakoid membrane of the chloroplast. The light-driven reactions include electron promotions for P-680 and P-700. $NADP^+$ is the electron carrier.

25. B is correct.

Calvin cycle (i.e., C_3 cycle) is a series of biochemical redox reactions in the stroma of chloroplasts in photosynthetic organisms.

It is one of the light-independent reactions used for carbon fixation.

RuBisCO is the key enzyme of the cycle.

Photosynthesis occurs in two stages.

1) *Light-dependent reactions* capture the energy of light and use it to make energy storage and transport molecules ATP and NADPH.

2) *Light-independent reactions* (Calvin cycle) use energy from short-lived electronically excited electron carriers to convert carbon dioxide and water into organic compounds used by the organism (and animals that feed on it).

This set of reactions is *carbon fixation.*

continued...

 Copyright © Sterling Education

Calvin cycle has four steps:

1) carbon fixation,
2) reduction phase,
3) carbohydrate formation, and
4) regeneration phase.

ATP and NADPH provide energy for the chemical reactions in this sugar-generating process.

Sunlight provides the energy for the reaction:

ATP + NADPH + carbon dioxide (CO_2) → sugar

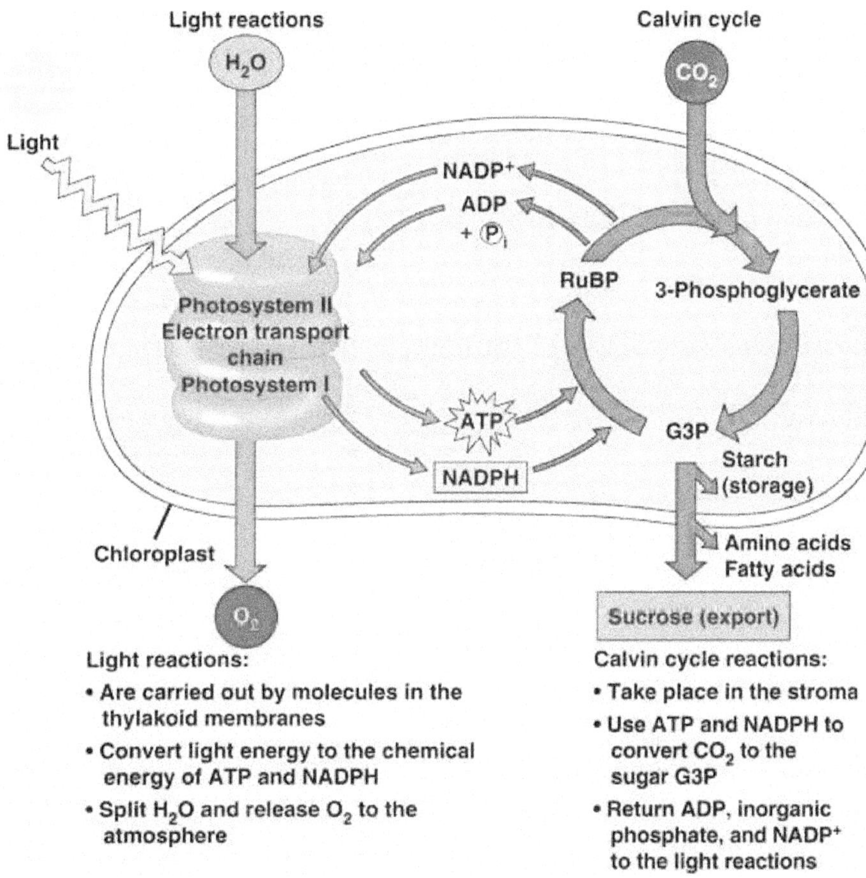

Products of light-dependent (ATP and NADPH) and light-independent reactions (sugars)

26. E is correct.

Electrons flow during photosynthesis in H_2O → $NADP^+$ → Calvin cycle.

27. D is correct.

Light-dependent reactions of photosynthesis produce NADPH and ATP.

28. D is correct.

Calvin cycle occurs in the *stroma* of chloroplasts.

29. C is correct.

Blue-colored plants reflect light of blue wavelength (380 – 500 nm); visible light spectrum (380 – 700 nm).

30. A is correct.

Three components of an ATP molecule are adenosine (*base*), ribose (*sugar*), and three *phosphate* groups.

31. C is correct.

Wavelength *reflected* determines the color of an object.

32. A is correct.

Energy is *released* by the hydrolysis of ATP (triphosphate) into ADP (diphosphate).

33. E is correct.

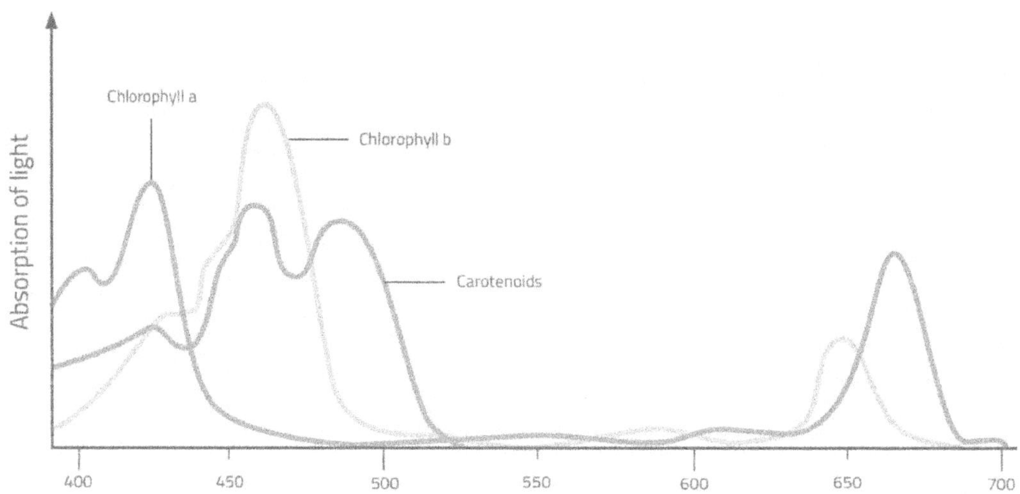

Photopigments absorb specific wavelengths in a visible light spectrum from 380 to 700 nm

Copyright © Sterling Education

34. C is correct.

Autotrophs (i.e., *producers*) synthesize complex organic compounds (i.e., carbohydrates, fats, proteins) from simple substances in their environment.

Autotrophs use energy from light (i.e., photosynthesis) or inorganic chemical reactions (i.e., chemosynthesis).

Autotrophs are the producers in a food chain (plants on land and algae in water), in contrast to heterotrophs that consume autotrophs.

Autotrophs can make their food and do not need living energy or an organic carbon source.

Autotrophs reduce carbon dioxide (CO_2) to make organic compounds.

Most autotrophs use water to reduce, but some can use other hydrogen compounds.

Phototrophs (i.e., green plants and algae), a type of autotroph, convert physical energy from sunlight into chemical energy in the form of reduced carbon.

35. A is correct.

Carotenoids reflect light with yellow, red, and orange wavelengths.

36. D is correct.

Mushrooms are *heterotrophs* and derive nutrients from *complex organic molecules*.

37. E is correct.

Plants use chlorophyll *a*, chlorophyll *b*, and carotenoid accessory pigments to absorb light over a broad spectrum of wavelengths.

38. B is correct.

The *sun* is the *ultimate energy source* for living organisms.

39. A is correct.

Photons are discrete packets of light.

40. B is correct.

Most plants appear green because chlorophylls do *not* absorb wavelengths of green light (500 – 560 nm).

41. A is correct.

Electromagnetic spectrum (ROY G BIV) has higher energy light with shorter wavelengths.

Any duplication (copies, uploads, PDFs) is illegal.

42. B is correct.

Thylakoids are membrane-bound compartments inside chloroplasts and cyanobacteria and the site of photosynthesis.

43. B is correct.

Energy from photons is transferred to electrons, exciting them to a higher energy level.

44. C is correct.

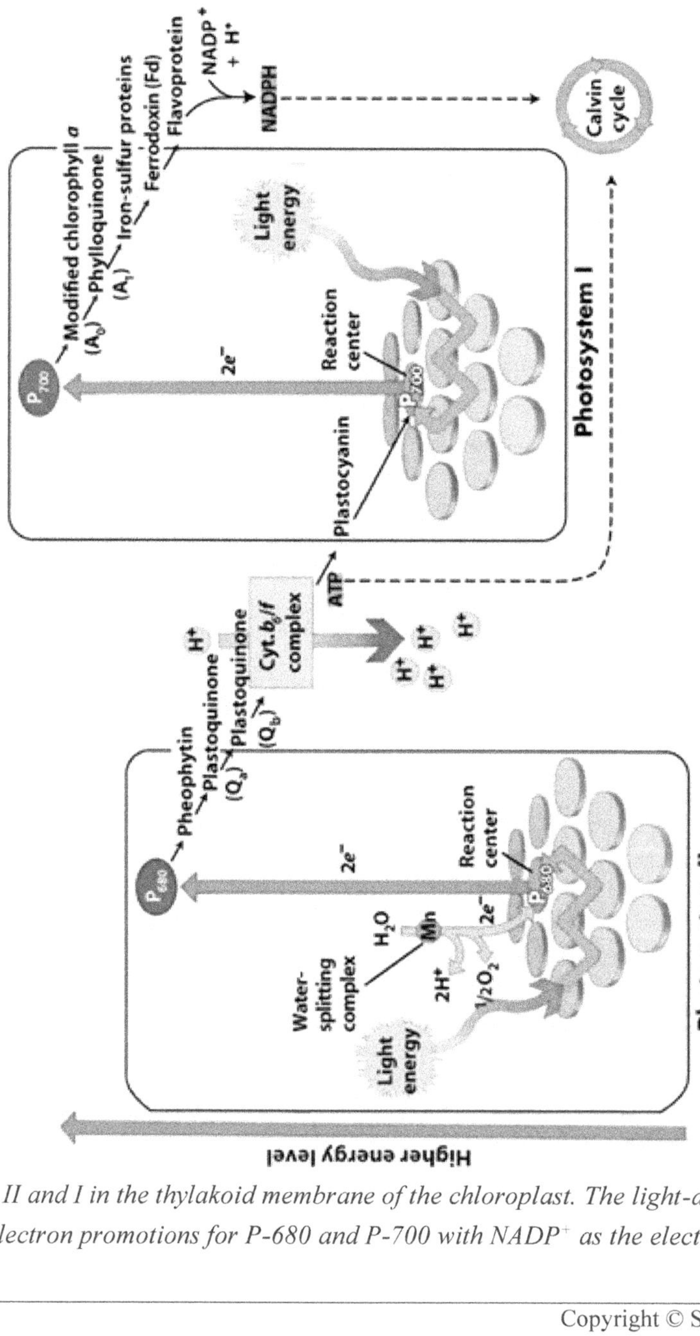

Photosystem II and I in the thylakoid membrane of the chloroplast. The light-driven reactions include electron promotions for P-680 and P-700 with $NADP^+$ as the electron carrier.

Copyright © Sterling Education

45. E is correct.

Molecules can *release energy* such as fluorescence, heat, or light.

46. A is correct.

Photosynthesis reaction:

$$6\ CO_2 + 6\ H_2O \rightarrow C_6H_{12}O_6 + 6\ O_2$$

Reactants are 6 carbon dioxide (CO_2) and 6 water (H_2O) molecules, converted by light energy captured by chlorophyll pigments into 1 sugar ($C_6H_{12}O_6$) and 6 oxygen (O_2) molecules.

47. D is correct.

Photosystems are in the *thylakoid* membrane of chloroplasts.

48. A is correct.

Light-dependent reactions occur within *thylakoid membranes.*

49. E is correct.

Chlorophyll *a* is a form of chlorophyll used in photosynthesis, absorbing most energy from wavelengths of violet-blue and orange-red light.

Chlorophyll *a* reflects green and yellow and contributes to the green color of many plants.

Photosynthetic pigment is essential for photosynthesis in eukaryotes and cyanobacteria because it is a primary electron donor in the electron transport chain.

50. C is correct.

Products of the *light-dependent reaction* are ATP, NADPH, and O_2 gas.

51. A is correct.

Calvin cycle has four steps:

carbon fixation,

reduction phase,

carbohydrate formation, and

regeneration.

ATP and NADPH provide energy for the chemical reactions in this *sugar-generating process.*

Sunlight provides energy for the reaction:

$$\text{ATP} + \text{NADPH} + \text{carbon dioxide } (CO_2) \rightarrow \text{sugar}$$

Any duplication (copies, uploads, PDFs) is illegal.

52. B is correct.

Calvin cycle (i.e., C_3 cycle) is a series of biochemical redox reactions in the stroma of chloroplasts in photosynthetic organisms.

Calvin cycle is one of the light-independent reactions used for carbon fixation.

RuBisCO is the key enzyme of the Calvin cycle.

Photosynthesis occurs in two stages in a cell.

Light-dependent reaction captures energy from light and uses it to make the energy storage and transport molecules ATP and NADPH.

*Light-independen*t Calvin cycle uses energy from short-lived electronically excited electron carriers to convert carbon dioxide and water into organic compounds used by the organism (and animals that feed on it).

This set of reactions is *carbon fixation*.

53. D is correct.

Water is the source of *oxygen* and electrons released by a photosystem.

54. C is correct.

Light-independent reaction of photosynthesis used for carbon fixation is the *Calvin cycle*.

55. B is correct.

Water is the source of oxygen and *electrons* released by a photosystem.

56. E is correct.

Electron transport chain occurs in light reactions of photosynthesis (in chloroplasts) and *cellular respiration* (in mitochondria).

57. C is correct.

H^+ gradient is formed across the *thylakoid membranes* of chloroplasts during photosynthesis.

 Copyright © Sterling Education

Notes for active learning

Notes for active learning

Copyright © Sterling Education

Annotated Glossary of Botany Terms

A

ab (*prefix*) – meaning *from, away from, or outside.*

Abaxial – the surface of an organ facing away from the organ's axis (e.g., the lower surface of a lateral organ such as a leaf or petal. Contrast *adaxial.*

Abort – to abandon the development of a structure or organ.

Abscission – the shedding of an organ that is mature or aged (e.g., ripe fruit or an old leaf).

Abscission zone – a specialized layer of tissue that allows an organ to be shed by abscission when it is ripe or senescent. Such tissue is commonly formed, for example, at the base of a *petiole* or *pedicel.*

Absorption zone – root: zone of root hairs.

Acaulescent – having no apparent stem, or at least none visible above the ground surface (e.g., some species of *Agave*, *Oxalis*, and *Attalea*. Antonym of *caulescent* (i.e., possessing stem)

Accessory pigment – absorbs light energy (e.g., individual colors) that chlorophyll *a* cannot absorb; the accessory pigments enable plants to capture more energy in light.

Accrescent – increasing in size with age, such as a calyx that grows after the corolla has fallen (e.g., *Physalis peruviana*).

Accumbent – lying against another part of the plant. When applied to a cotyledon, one edge of the cotyledon lies along the radicle.

Achene – a dry, one-seeded indehiscent fruit in which the true fruit is not the so-called "*berry*," but the achenes, the "*seeds*" on the infructescence. One-seeded indehiscent dry fruit of Compositae, cypsela.

Acicular slender or needle-shaped.

Acropetal – moving from roots to leaves (e.g., molecular signals in plants.

Acrophyll – the regular leaves of a mature plant, produced above the base instead of bathyphyll.

Acrostichoid (*describing a type of sorus*) – covering the entire abaxial surface of a frond, usually densely so, as in *Elaphoglossum* and *Acrostichum.*

Actino- (*prefix*) – indicates a radial pattern, form, or morphology.

Actinodromous (*leaf venation*) – palmate or radially arranged venation with three or more primary veins arising at or near the base of the leaf and either reaching the margin or not.

Actinomorphic – regular or radially symmetrical; may be bisected into similar halves in at least two planes. It applies to steles and flowers in which the perianth segments within each whorl are alike in size and shape. Compare *regular*; contrast *asymmetrical*, *irregular*, and *zygomorphic*.

Any duplication (copies, uploads, PDFs) is illegal.

Aculeate – with prickles (e.g., the stem of a rose).

Acuminate – tapering gradually to a point. Contrast *acute* and *mucronate.*

Acute – 1. sharply pointed but not drawn out. Contrast *accuminate.* 2. converging at an angle of less than 90°. Contrast *obtuse.*

Ad- (*prefix*) – meaning *near or toward*; also meaning *added to.*

Adaptation – a trait or characteristic of an organism that makes it more likely to survive in its environment.

Adaxial – the surface of an organ facing toward the organ's axis (e.g., the upper surface of a lateral organ such as a leaf or petal). Contrast *abaxial.*

Adelphia (pl: *adelphiae*) – a bundle or structure of stamens forms one unit in an adelphous flower, such as the stamen tube around the pistil of Hibiscus.

Adelphous – having organs, particularly filaments such as stamens, connected into one or more adelphiae, whether in bunches or tubes, commonly seen in families such as *Malvaceae*. The term's usage is inconsistent; some authors include closely bunched filaments, while others include only adelphiae, in which filaments are connected at least at their bases.

Adenosine triphosphate (ATP) – combining adenosine diphosphate (ADP) and phosphate. *ATP is the primary energy currency of cells.*

Adherent – slightly united to an organ of another kind, usually to a part of another whorl (e.g., a sepal connected to a petal). Contrast *adnate.*

Adnate – grown from or closely fused to an organ of a different kind, especially along a margin (e.g., a stamen fused to a petal). Adnate anthers have their halves attached to the filament through most of their length. Contrast *connate.*

Adventitious – roots originate from the stem and are produced in an unpredictable or unusual position (e.g., an adventitious bud produced from a stem rather than a more typical leaf axil). Adventitious roots may develop from nodes of prostrate stems of some plant species or the hypocotyl rather than from the radicle of a germinating monocotyledon.

Adventive – introduced accidentally (usually referring to a weed).

Aerial – of the air; growing or borne above the ground surface or water.

Aerobic respiration – processes that require oxygen are *aerobic.*

Aestivation – the arrangement of sepals and petals or their lobes in an unexpanded flower bud. Contrast *vernation.*

Aggregate fruit – a cluster of fruits formed from the free carpels of a single flower (e.g., a blackberry). Compare *multiple fruits.*

Agricultural weed – see *weed.*

Copyright © Sterling Education

Alate – having a wing or wings.

Albumen – an older name for the endosperm of flowering plants. Except for being a storage tissue for nutrients, it is not like the albumen (i.e., egg white) of animal embryos.

Albuminous (*of seeds*) – containing endosperm.

Alcoholic fermentation – occurs in some plants and unicellular organisms such as yeast and bacteria. The process converts pyruvic acid into ethyl alcohol and a carrier compound (giving off CO_2), which allows glycolysis to continue.

Alien – any plant introduced to an area outside its natural range. Often used interchangeably or in combination with *foreign*, *exotic*, *non-native*, and *non-indigenous*.

Alkaloid – any loosely defined class of organic compounds found in the tissues of many species of plants. Alkaloid molecules have one or more alkaline-reacting nitrogen atoms in their carbon structures. Many alkaloids are commercially important as drugs or poisons (e.g., caffeine, morphine, quinine, and strychnine, each occurring naturally in certain plants).

Alternate – 1. (*of leaves or flowers*) borne singly at distinct levels along a stem, including spiraled parts. Contrast *opposite*. 2. occurring between something else (e.g., stamens alternating with petals).

Alternate leaf arrangement (or *spiral leaf arrangement*) – one leaf per node.

Ament – a synonym of *catkin*.

Amphitropous – when the ovule is bent so that both ends are near. Contrast *anatropous*, *campylotropous*, and *orthotropous*.

Amplexicaul – with the base dilated and clasping the stem, usually leaves.

Amylum star – a vegetative propagative body filled with starch (amylum) located around the lower nodes of certain stoneworts.

Anaerobic respiration – processes that do not require oxygen are "*anaerobic*" processes.

Anastomose – branching and then rejoining, as with leaf venation.

Anastomosis – a connection or fusion of veins that are typically diverging or branching, forming a network.

Anatropous (*of an ovule*) – inverted so that the micropyle faces the placenta (the most common ovule orientation in flowering plants). Contrast *amphitropous*, *campylotropous*, and *orthotropous*.

Androecium – collective term for stamens. *Stamen* (filament and anther), *anther* (structure containing pollen grains), and *filament* (structure connecting anther to a receptacle).

Androdioecious – having bisexual flowers and male flowers on separate individuals. Contrast *andromonoecious*, *polygamodioecious*, *polygamomonoecious*, and *polygamous*.

Any duplication (copies, uploads, PDFs) is illegal.

Androecium – a collective name for the male reproductive parts of a flower; the stamens of a flower considered collectively. Contrast *gynoecium*. Abbreviated *A*; (e.g., *A 3+3* indicates six stamens in two whorls).

Androgynophore – a stalk bearing both the androecium and gynoecium of a flower above the level of insertion of the perianth.

Androgynous – having male and female flowers in the same inflorescence.

Androphore – the stalk or column supporting the stamens, particularly flowers.

Andromonoecious – having bisexual flowers and male flowers on the same individual plant. Contrast *androdioecious*, *gynomonoecious*, *polygamodioecious*, *polygamomonoecious*, and *polygamous*.

Anemophilous – adapted to pollination by wind.

Anemophily – adaptation to pollination by wind.

Angiosperm – a flowering plant with developing seeds enclosed in an ovary.

Anisomery – having a floral whorl with a different (usually smaller) number of parts from the other floral whorls.

Anisotomic – branching, with branches having unequal diameters, like a trunk and its branch. Contrast *isotomic*.

Annual – a plant that completes its life cycle (i.e., germinates, reproduces, and dies) within a year or growing season.

Annulus – 1. a ring-like structure in the form of a ring. Pappus bristles are sometimes attached to a ring called an *annulus* or *disk* at the top of the achene beak. In some pollen grains, the exine around the apertures is either thicker or thinner. In pores, this border is termed an *annulus*. Certain flowers have ring-like constrictions at the mouth of the flower. 2. a ring of specialized cells on the sporangium.

Anomalous – secondary growth when there are multiple, short-lived layers of cambium.

Anterior – positioned in front of, toward the apex. Compare *distal*.

Anthemoid (or *senecioid*) (*in Compositae*) – a style with a brush-like tuft of sweeping hairs at the tip of each style branch.

Anther – the pollen sac on a male flower; the pollen-bearing part of a stamen.

Antherode – a sterile anther of a staminode.

Anthesis – 1. (*of a flower*) the period during which pollen is presented, or the stigma is receptive. 2. (*of a flowering plant*) the period during which flowers in anthesis are present. It is not defined for some cases, such as when pollen is released in the bud.

Anthocarp – fruit in which some part of the flower persists attached to the pericarp (e.g., in *Nyctaginaceae*).

 Copyright © Sterling Education

Anthophore – a stalk-like structure, internode between the calyx and the other parts of the flower.

Anticlinal – pointing up, away from, or perpendicular to a surface. Contrast *periclinal.*

Antrorse – directed toward or upward (e.g., hairs on a stem). Contrast *retrorse.*

Apetalous – lacking petals.

Apex (pl. *apices*) – the tip, the point furthest from the attachment point.

Aphananthous (*of flowers*) – inconspicuous or unshowy, as opposed to *phaneranthous* or showy.

Aphlebia – imperfect or irregular leaf endings commonly found on ferns and fossils of ferns from the Carboniferous Period.

Apical – at or on the apex of a structure, usually a shoot, a stem, or a tree trunk (e.g., an apical meristem or an apical bud).

Apical dominance – the tendency of an apical bud to produce hormones that suppress the growth of buds below it on the stem.

Apical meristem (or *growing tip*) – an undifferentiated meristematic tissue in the buds and growing tips of plant roots. Its primary function is to trigger the growth of new cells in young seedlings at the tips of roots and shoots and forming buds. See *root apical meristem* (RAM) and *stem apical meristem* (SAM).

Apiculate – especially of leaves, ending in a short triangular point.

Apiphily – pollination whereby honeybees distribute pollen.

Apo- (*prefix*) – meaning "*away from*, *separate*, or *without.*

Apocarpous (*of a gynoecium*) – consisting of one or more carpels free from one another (or almost so).

Apogamy apomixis – when an embryo develops from unfertilized gamete, parthenogenesis.

Apomixis – asexual reproduction via seeds or spores in which embryos develop without fertilization, such that the genetic material they contain is a clone of the parent's genetic material. Plants produced are called *apomict.*

Apomorphy – in cladistics, a *different form* from the form of an ancestor (i.e., an innovation) of use in determining membership in a clade.

Apopetalous – having separate petals, not fused (*sympetalous*).

Apophysis – 1. the external part of a cone scale. 2. an outgrowth of an organ or an enlargement of a stem.

Apoplast – the extracellular space outside plant cell membranes, especially the fluid-filled cell walls of adjacent cells where water and dissolved material can freely flow and diffuse. *Fluid and material flows occurring in any extracellular space are apoplastic flow or transport.*

Any duplication (copies, uploads, PDFs) is illegal.

Apospory – the development of 2N gametophytes, without meiosis and spores, from vegetative or nonreproductive cells of the sporophyte.

Appendage – a secondary part attached to a main structure; an external growth that seldom has any obvious function, hence appendiculate.

Appendiculate – having the nature of or bearing appendages.

Appressed – pressed closely but not fused (e.g., leaves against a stem).

Aquatic plant – a plant whose natural habitat is water, living in or on water for a substantial part of its lifespan, generally restricted to fresh or inland waters.

Arachnoid – cobwebby, from being covered with fine white hairs.

Arboreal – of or relating to or formed by trees.

Arborescent – tree-like in growth or general appearance.

Arboretum – ***a*** taxonomically arranged collection of trees.

Archaeophyte – a non-native plant in a particular geographic area for some time. Contrast *neophyte*.

Archegonium – a multicellular structure or organ of the gametophyte phase of certain plants, producing and containing the ovum or female gamete. The corresponding male organ is called the *antheridium*.

Arctotoid – in the Compositae, a style with a ring of sweeping hairs borne on the shaft of the style proximal to the style branches.

Areolate – having or being composed of areoles, as an areolate crustose lichen.

Areole – a space between the threads of a net (e.g., that part of a leaf surface defined by each of the elements of a vein network), as with cacti, the area between the veinlets of a leaf. 2. a structure on a cactus's stem node, the cactus's region upon which spines and flowers are borne. 3. (*in lichenology*) a polygonal piece of a thallus surface where a crustose lichen is broken up like old dried and cracked paint or the polygonal "islands" of dried mud in a dry lakebed.

Aril – a membranous or fleshy appendage formed by expansion of the funicle that partly or wholly covers a seed (e.g., the fleshy outer layer of lychee fruit or in members of *Sapindaceae*).

Aristate – with a stiff, bristle-like awn or tip.

Article – a segment of a jointed stem or a fruit with constrictions between the seeds; an organ part that separates easily from the rest of the organ at a joint or articulation.

Articulate – jointed; separating freely, leaving a clean scar (e.g., the fronds of certain ferns where they join the rhizome).

Ascending – 1. (*of a stem*) spreading horizontally, then directed upward; an ascending stem is prostrate near its base, then erect. 2. (*of an ovule*) attached somewhat above the base.

 Copyright © Sterling Education

Asexual reproduction – reproduction that does not involve gametes. Often used interchangeably with vegetative reproduction.

Asperulous – having a rough, sandpapery texture (e.g., some leaf surfaces).

Asymmetrical – irregular or unequal; lacking any plane of symmetry.

Atactostele – vascular bundles dispersed.

Attenuate – narrowing gradually.

Auricle – an ear-shaped lobe, particularly a small, roundish, lateral appendage of a leaf or leaf-like organ.

Autotrophs (or *producers*) – use photosynthesis to convert the energy of sunlight, carbon dioxide (CO_2), and water into chemical energy (e.g., ATP) or food (glucose; $C_6H_{12}O_6$).

Awn – 1. any long, bristle-like appendage. 2. (*in Poaceae*) an appendage terminating or on the back of glumes or lemmas of some grass spikelets. 3. (*in Geraniaceae*) the part of the style remains attached to the carpel and separates from the carpophore (column). 4. a generally straight, stiff pappus element, varying from stiffly bristle-like to hard and needle-like. In *Strophanthus*, the awn is the beak of the seed stipe of the coma hairs.

Axils – an apical bud sprouting from the apex of the shoot of *Searsia angustifolia.* The axillary buds are dormant in scaly cataphylls, but the apical bud is bare, exposing emerging leaf rudiments.

Axil – where a leaf joins a stem. *The upper angle between one part of a plant and another (e.g., the stem and a leaf).*

Axile – on the axis of a placenta, on the central axis of the ovary.

Axillary – borne in or arising from the axil, usually referring to the axil of a leaf.

Axis – the main stem of a whole plant or inflorescence; also, the line along which this stem extends.

B

Baculiform – rod-like; longer than wide. Compare *cylindrical.*

Barb – a rear-facing point, as in a fishhook.

Barbellate – having barbed hairs (*barbellae*).

Bark – the tree's outer layer that protects against fire and insects while insulating from extreme heat or cold. The protective external layer of tissue on the stems and roots of woody trees and shrubs includes all living and non-living tissue external to the cambium.

Basal – situated or attached at the base.

Any duplication (copies, uploads, PDFs) is illegal.

Basifixed – something attached by its base (e.g., an anther attached to the filament). Compare *dorsifixed.*

Basipetal – 1. developing sequentially from the apex toward the base (i.e., with the youngest toward the base). For example, flowers in an inflorescence. 2. moving from leaves to roots (e.g., of molecular signals in plants).

Bathyphyll – a specialized leaf produced at the base of a plant, usually when the plant is immature, anchors the plant to a substrate, especially notable in the fern *Teratophyllum*. Contrast *acrophyll.*

Beak – a prominent, pointed terminal projection, especially of a carpel or fruit.

Berry – a type of indehiscent fruit with seeds immersed in the pulp (e.g., a tomato).

Bi- (*prefix*) – meaning *two* (e.g., bisulcate, having two sulci or grooves).

Biennial – a plant that completes its life cycle (i.e., germinates, reproduces, and dies) within two years or growing seasons. Biennial plants usually form a basal rosette of leaves in the first year and then flower and fruit in the second year.

Bifid – forked; cut in two for about half its length. Compare *trifid.*

Bifoliate (*of a compound leaf*) – having precisely two leaflets, usually in a symmetrical pair. Compare jugate lobed leaf (e.g., most species of *Bauhinia*).

Bifusiform – fusiform with a pinch in the middle.

Biternate – ternate, with each division divided into three.

Bilabiate – having two lips (e.g., the form of the petals in many irregular flowers).

Bilateral – 1. having two distinguishable sides, such as the two faces of a dorsiventral leaf. 2. arranged on opposite sides (e.g., leaves on a stem; cf. distichous and opposite). 3. bilaterally symmetrical, as in a leaf with a symmetrical outline.

Biloculate – having two loculi (e.g., in *anthers* or *ovaries*).

Binomial – using names consisting of two words to form the scientific name (or combination) in a Latin form. *For example, the first is the name of the species-genus, and the second is the specific epithet given to that species to distinguish it from others in the same genus.*

Binomial nomenclature – the system of nomenclature in which the scientific name of a species (and not of a taxon at any other rank) is a combination of two names, the first name being the generic name. The second name is referred to botanically as the specific epithet. Note that the two names (not just the second name) constitute the species name.

Bipinnate – doubly pinnate (e.g., a compound leaf with individual leaflets pinnately divided).

Bipinnatisect – a pinnatisect leaf with deeply dissected segments.

Bipolar plant body – both root and shoot systems present.

 Copyright © Sterling Education

Bitegmic (*of an ovule*) – covered by two integuments. Contrast *unitegmic*.

Bivalve – having two valves or hinged parts. Contrast *trivalve*.

Blade – the lamina or flattened leaf part, excluding the stalk or petiole.

Bloom – a fine white or bluish waxy powder on plant parts, usually stems, leaves, and fruits. It is easily removed by rubbing.

Bole – the trunk of a tree, usually the portion below the lowest branch. Compare *canopy*.

Bostrychoid – arranged on a conical surface (like a snail shell); used to describe inflorescences in which the buds are arranged almost helically on the outside of a long, tapering, conical rachis.

Botany – the scientific study of plants and plant-like organisms.

Bough – any of the larger branches of a tree. *Bough* is another word for *branch*, below, though it usually refers to bigger ones connected directly to the trunk.

Brachyblasts – shortened shoots of pines, larches, and some other *Pinaceae* conifers.

Bract – a modified leaf associated with a flower or inflorescence differing in shape, size, or color from other leaves (without an axillary bud).

Bract scales – sterile bracts under seed scales in conifers.

Bracteate – possessing bracts.

Bracteole – a small bract borne singly or in pairs on the pedicel or calyx; synonymous with bractlet.

Bracteolate – possessing bracteoles (bractlets).

Bractlet – see *bracteole*.

Branch – a division of a stem arising from the main stem of a plant.

Branchlet – a small branch.

Brevideciduous – a plant that loses all its leaves only briefly before growing new ones so that it is leafless for only a short time (e.g., approximately two weeks).

Bristle – a straight, stiff hair (smooth or with minute teeth); the upper part of an awn (when the latter is bent and has a lower, stouter, and usually twisted part called the column).

Brochidodromous – pinnate leaf venation in which the secondary veins do not terminate at the leaf margin but are joined in a succession of prominent arcs.

Brochus (pl. *brochi*) – width of one lumen of a pollen grain reticulum and half of the width of the surrounding muri (walls), hence *heterobrochate* and *homobrochate*, where the lumina are of different or similar sizes, respectively.

Any duplication (copies, uploads, PDFs) is illegal.

Bryophyte – informally, any plant that is a moss, hornwort, or liverwort. Formally, these plants are placed in three separate divisions: *hornworts* (Anthocerophyta), *liverworts* (Marchantiophyta), and *mosses* (Bryophyta).

Buds – embryonic shoots.

Bulb – short, thick underground storage shoot with the prevalence of leaf tissues. A thick storage organ, usually underground, consisting of a stem and leaf bases (the inner ones fleshy).

Bulbel – a bulb arising from another bulb. See *bulblet.*

Bulbil – a small, deciduous bulb or tuber formed in the axil of a leaf or pinna (a means of vegetative propagation).

Bulblet – a bulb arising from another bulb.

Bullate – having rounded or globular blisters on the surface.

Burr – 1. a prickly fruit. 2. a rough or prickly propagule consisting of a seed or fruit and associated floral parts or bracts.

Buttress root – growing from an above-ground stem or trunk and providing support (e.g., commonly of *Ficus macrophylla*).

Byssoid – a growth form of a lichen thallus that is whispy, like teased wool.

C

Caducous – falling off early (e.g., the sepals of poppies, which fall off when the petals begin to open). Compare *persistent* and *fugacious.*

Caespitose– tufted or turf-like (e.g., the growth form of some grasses).

Calcarate – possessing a *spur.*

Calcareous – a soil or a lichen substrate rock type rich in or composed mainly of calcium carbonate.

Calciphytes – plants adapted to the over-presence of $CaCO_3$.

Callose – hardened; thickened, callous.

Callus (pl. *calli*) – 1. a protruding mass of tissue. 2. undifferentiated tissue growth formed in response to wounding may be grown in vitro. 3. (*in orchids*) fleshy outgrowths from the labellum can be shaped from papillae to plates. 4. (*in grasses*) a hardened extension from the base of a floret (formed from the rachilla joint or the base of the lemma) may or may not elongate and is often covered in hairs or bristles.

Calyciflorous – having petals and stamens attached to the calyx.

 Copyright © Sterling Education

Calycophyll – a leaf-like structure formed from a sepal or calyx lobe that enlarges, usually many-fold, before or after anthesis, especially when most of the other sepals or calyx lobes retain their original size. More extreme than an accrescent calyx, calycophylls are in *Rubiaceae*. Compare *semaphyll* and *pterophyll.*

Calyculate – having an epicalyx.

Calyculus – 1. a cup-shaped structure formed from bracts resembling an outer calyx. 2. in some *Asteraceae*, a circle of bracts below the involucre.

Calyptra – a hood or lid. See *operculum.*

Calyx (pl. *calyces*) – a collective term for the sepals of one flower; the outer whorl of a flower, usually green. Compare *corolla.*

Calyx tube – formed by the fusion of the sepals (calyx), at least at the base.

Cambium – the growing part of the trunk that becomes either xylem or phloem. A tissue layer that provides partially undifferentiated cells for plant growth. *A layer of growing tissue separates the xylem and phloem and continuously produces new xylem and phloem cells.*

Campanulate – bell-shaped.

Camptodromous – pinnate venation in which the secondary veins curve toward the margins, sometimes becoming parallel with them and not reconnecting with other veins to form loops.

Campylotropous – when the ovule is oriented transversely (i.e., with its axis at right angles to its stalk) and with a curved embryo sac. Compare *amphitropus*, *anatropous*, and *orthotropous.*

Canaliculate – channeled; having a longitudinal groove.

Canescent – approaching white, as in a leaf covered with white down or wool.

Canopy – 1. the branches and foliage of a tree; the crown. 2. the protective upper layer of a forest. Compare *trunk*

Capillary – 1. a tube, pore, or passage with a narrow, internal cross-section. 2. slender; hair-like.

Capitate –1. (*of an inflorescence*) Having a knob-like head, with the flowers unstalked and aggregated into a dense cluster. 2. (*of a stigma*) like the head of a pin.

Capitulum – a dense cluster of sessile or subsessile flowers or florets (e.g., a flower head in the daisy family, *Asteraceae*). See *pseudanthium.*

Capsule – a dry fruit formed from two or more united carpels and dehiscing when ripe (usually by splitting into pieces or opening at the summit by teeth or pores).

Cardioid – in *Asteraceae*, having a style with a ring of sweeping hairs borne on the shaft of the style below the style branches.

Any duplication (copies, uploads, PDFs) is illegal.

Carina – part of a papilionaceous flower consisting of two petals, commonly united, which encloses the organs of fructification. See *keel.*

Carinal canal – a longitudinal cavity in the stems of *Equisetum* and extinct *Equisetopsida*, coinciding with a ridge in the stem surface.

Carneous – flesh-colored, especially as applied to some flowers.

Carnose (or *carnous*) – fleshy or pulpy in texture, mainly applied to some tissues or organs. Contrast *coriaceous* and *corneous*.

Carotenoids – accessory pigments that include yellow, red, and orange pigments that color carrots, bananas, squash, flowers, and autumn leaves.

Carpel – structure enclosing ovules may correspond with locules or placentas. Carpel and pistil are equivalent when there is no fusion; if fusion occurs, then 2 or more carpels unite into one pistil. *The basic female reproductive organ in angiosperms consists of a single sporophyll or a single locule of a compound ovary with a style and a stigma. The gynoecium is the collective term for all the carpels of a single flower.*

Carpopodium – on achenes (Cypselas), an elongation of the base of the gynoecium which looks distinct; the abscission zone, where the achene is separated from the receptacle.

Cartilaginous – hard and tough; gristly. Compare *corneous* and *coriaceous*.

Caruncle – a small piece of flesh-like tissue, typically lumpy or warty, growing on the testa near the hilum. Contrast *aril.*

Caryopsis – a dry, indehiscent, one-seeded fruit in which the seed coat is closely fused to the fruit wall (e.g., in most grasses).

Casparian strip – part of endodermis cell walls that prevent *apoplastic* transport. A continuous band of suberin in the radial primary cell walls of the endodermis in vascular plant stems, and roots forms a permeability barrier to the passive diffusion of external water and solute into the vascular tissue.

Cassideous – hood-shaped, helmet-shaped, or bonnet-shaped; generally referring to floral anatomy (e.g., in the flowers of *Aconitum*, *Satyrium*, etc.).

Casual alien – an exotic plant that appears with no apparent human assistance but does not develop a sustained population or one that persists only by repeated new introductions. Compare *alien.*

Cataphyll – any plant structure that is morphologically a leaf but has, at most, an incidental or transient photosynthetic function. They are either shed when their main function has been completed or are incorporated into structures where, when dead, they serve a protective or supportive purpose.

Catenulate – in the shape of a chain; formed of parts or cells connected as if chained together (e.g., some diatoms, algae, and cyanobacteria such as *Anabaena*). See *concatenate*.

Catkin – a spike, usually pendulous, in which the primarily small flowers are unisexual and without a conspicuous perianth (e.g., in willows, poplars, oaks, and casuarinas). The individual flowers often have scaly bracts and are generally wind-pollinated. *Catkins are usually shed as a unit.*

 Copyright © Sterling Education

Caudate – having a narrow, tail-like appendage or tip (e.g., a drip tip). Contrast *acuminate*, *cuspidate*, and *mucronate*.

Caudex (pl. *caudices*) – 1. the stem of a plant, especially a woody one. 2. a rootstock, particularly a basal stem structure or storage organ from which new growth arises. Compare *lignotuber.*

Caudiciform – stem-like or caudex-like; sometimes used to mean *thick-stemmed.*

Caulescent – possessing a well-developed stem above ground, similar to cauline. Antonym of *acaulescent* (i.e., lacking an apparent stem).

Cauline – borne on an aerial stem or caulis, as with leaves, flowers, or fruits (when applied to the latter two organs, usually referring to older stems; synonymous with cauliflorous).

Cell – 1. the basic, microscopic unit of plant structure, consisting of compartments in a viscous fluid surrounded by a cell wall. 2. a cavity of an anther or ovary.

Cellular respiration – heterotrophs convert stored chemical energy into usable ATP energy. *Cellular respiration uses oxygen in the mitochondria to chemically break down organic molecules such as glucose and release the energy stored in its bonds.* Cellular respiration equation: $C_6H_{12}O_6 + 6\ O_2 \rightarrow 6\ CO_2 + 6\ H_2O$ (+ ATP energy)

Cenanthous (*of a perianth*) – lacking both stamens and pistil (i.e., a flower with neither androecium nor gynoecium).

Central cell – the biggest cell of the embryo sac, with two (or sometimes one) haploid (1N) nuclei.

Centrifixed –a two-branched organ attached by its center (e.g., a hair or anther).

Ceraceous – having a waxy appearance, color, or texture (e.g., flowers of many species of *Ceropegia* and the waxy fruit of some species of *Myrica*).

Cernuous – nodding, falling headlong or face down; inclined, stooping, or bowing forward. Applied to many species with a nodding, stooping habit, such as many *Narcissus* and *Dierama* species. Many plant species bear the specific epithet *cernua*.

Cespitose (or *caespitose*) – tufted or turf-like (e.g., the growth form of some grasses)

Chartaceous – having a papery texture.

Chamber – a cavity of an ovary.

Channeled – sunken below the surface, resulting in a rounded channel.

Chasmogamous – of flowers that are pollinated when the perianth is open. Compare *cleistogamous*.

Chasmophyte – a plant adapted to growing in crevices or hollows, such as in cliff faces. Compare *cremnophyte*.

Chemosynthesis – a process by which organisms use inorganic compounds as their energy supply.

Any duplication (copies, uploads, PDFs) is illegal.

Chimera – composed of two or more genetically distinct tissues, most commonly because of a graft and sometimes by mutations that occur during cell division or cellular transfers during seed development.

Chiropterophilous – pollinated by bats.

Chlorophyll – the green pigment in leaves responsible for trapping light energy from the sun. A chemical inside of leaves that makes them green and helps in photosynthesis. Any of a variety of chemical pigments in chloroplasts that are essential for photosynthesis.

Chlorophyll *a* – photopigment directly involved in the light reactions of photosynthesis.

Chlorophyll *b* – assists chlorophyll *a* capture light energy.

Chloroplast – a specialized component of specific cells; contains chlorophyll and is responsible for photosynthesis. An organelle present in plant cells that contains chlorophyll.

Chlorosis – an abnormal lack or paleness of color in a normally green organ.

Cilia (sing. *cilium;* adj. *ciliate*) – tiny hairs or hair-like protrusions confined to the margins of an organ; in motile cells, minute, hair-like protrusions which aid motility.

Circinate – spirally coiled with the tip innermost (e.g., circinate vernation of the developing fronds of most ferns).

Cirrhose (*of a leaf*) – ending in a tendril at the apex.

Citric Acid Cycle (or *Krebs Cycle*) – occurs in mitochondria after glycolysis and is the second of three phases of cellular respiration; it produces 2 ATP molecules, 10 energy carrier molecules, and CO_2 from each glucose molecule.

Cladode – a photosynthetic branch or stem, often leaf-like and usually with foliage leaves either absent or much reduced. Compare *phyllode.*

Class – the principal category for taxa ranking between division and order.

Clathrate – shaped like a net or lattice; pierced with apertures, as with a cage.

Clavate – club-shaped.

Clavuncula – in *Apocynaceae*, an enlarged, drum-shaped stigma of which the sides and lower surface are the receptive zones. Coherent with the anthers or not.

Claw – 1. a narrow, stalk-like, basal portion of a petal, sepal, or bract. 2. (in *Melaleuca*) the united portion of a stamen bundle.

Cleistogamous – having flowers that self-pollinate and never open fully or which self-pollinate before opening. Compare *chasmogamous.*

Climber – a plant growing erect by leaning on or twining around another structure for support or clinging with tendrils.

 Copyright © Sterling Education

Cline (adj. *clinal*) – a continuous morphological variation in form within a species or sometimes between two species.

Clone – a plant derived from the asexual vegetative reproduction of a parent plant, with both plants having identical genetic compositions.

Coalescent – having plant parts fused or grown together to form a single unit.

Cochleariform – concave and spoon-shaped.

Cochleate – coiled like a snail's shell.

Coenobium – an arranged colony of algae that acts like a single organism.

Coenocyte – a single cell with multiple nuclei formed when nuclear division was not followed by cytokinesis.

Coleoptile – one type of sheath in the structure of monocotyledonous seeds. A protective sheath or cap (*pileus*), generally more or less pointed, covers the monocotyledonous plumule as it emerges from the soil. *It generally turns green and contributes to photosynthesis until the primary growth of the seedling supersedes its function.* Contrast *coleorhiza*, which remains underground until it is superseded as the roots emerge.

Coleorhiza – a sheath in the structure of monocotyledonous seeds. Coleorhiza connects the coleoptile to the radicle and protects the monocotyledonous radicle during germination. Unlike the coleoptile, the coleorhiza is associated with the root and does not emerge from the soil during germination. Contrast *coleoptile*.

Collenchyma – – living supportive tissue. A specialized tissue consisting of living cells with unevenly thickened cellulose and pectin cell walls supports organs such as leaves and young stems composed of primary plant tissues.

Colleter – a multicellular, glandular hair that usually produces a mucilaginous substance located on sepals, stipules, petioles, or nearby parts of stems; commonly found on plants in the order Gentianales.

Columella – in flowering plants, the central axis of the cone or fruit (e.g., in *Callitris*).

Column – 1. a structure extending above the ovary and incorporating the style and stamens, also known as the *gynostegium* (e.g., in orchids and milkweeds). 2. (*in grasses*) the lower, stouter, and usually twisted part of an awn, distinct from the slender upper part or bristle.

Columnar – shaped like a column.

Coma – 1. a tuft of hairs from testa or funiculus at one or both ends of some seeds (e.g., in *Strophanthus*, *Asclepias*, or *Alstonia*). 2. sterile bracts (e.g., in *Curcuma*, *Ananas*, or *Eucomis*). 3. sterile flowers (e.g., in *Muscari* and *Leopoldia*, at the apex of some inflorescences. 4. a tuft of hairs at the base of some flowers (e.g., in *Pfaffia gnaphalioides*). 5. a tuft of hairs at the apex or base of some spikelets. 6. an axil tuft of hairs in inflorescences in some *Poaceae* (e.g., in *Eragrostis comata*).

Any duplication (copies, uploads, PDFs) is illegal.

Commercial name – a name often of no botanical standing and not governed by the ICNCP. The term generally applies to names such as trademark names, names covered by plant breeders' rights, patents, and promotional names, which are often used to enhance the sale of a plant.

Commissure – the seam or face at which two carpels adhere. See *fissure* and *suture*.

Community – an ecological assemblage of plants that characteristically occur together.

Companion cells – nucleate "helpers" to anucleate sieve tube cells.

Complex tissues – with more than one type of cell.

Compound – composed of several parts. For example, a leaf composed of multiple leaflets, a gynoecium composed of multiple carpels, or an inflorescence comprised of multiple smaller inflorescences.

Compound fruit – originated from the whole inflorescence.

Compound leaves – with two or more levels of hierarchy.

Compound palmate – having leaflets that radiate from a central point (usually at the top of a petiole), like spread-out fingers radiating from the palm of a hand. Compare *palmate*.

Compressed – flattened lengthwise, either laterally (from side to side) or dorsally (from front to back).

Concatenate – joined together in a chain-like form. See *concatenate* and *catenate*.

Concolorous – having the same color throughout; uniformly colored.

Conduplicate – arranged such that two sides of a flat surface are folded along the midline to face each other. See *ptyxis*, *aestivation*, and *vernation*.

Cone – tapering mass of ovule- or spore-bearing scales or bracts. Fruit, usually woody, ovoid to globular, including scales, bracts, or bracteoles arranged around a central axis (e.g., in gymnosperms, especially conifers and Casuarina).

Conflorescence – a rarely used term describing substantial differences between the overall structure of an inflorescence and that of its branches (e.g., the bottlebrush multiple-flower head of members of the genus *Callistemon*).

Conifer – a type of tree or shrub-bearing cone. Coniferous trees, like pines, grow their seeds inside cones. They also do not drop their needles in the winter and tend to have much softer, more bendable wood than deciduous trees. That is because keeping their needles means that their branches need to bend under the weight of heavy snow.

Connate – fused to another organ (or organs) of the same kind (e.g., petals in a gamopetalous corolla tube). Compare *adnate*.

Connective – the part of an anther that connects the anther cells.

Connivant – coming into contact or converging.

 Copyright © Sterling Education

Conspecific – belonging to the same species.

Contiguous – adjoining, touching, but not united.

Contort (*of sepals or petals*) – imbricate aestivation in which one side of each segment overlaps one of the adjacent segments, and the other side is overlapped by the other adjacent segment. See *convolute.*

Contorted – twisted out of the normal shape.

Contractile roots – which pull the plant deeper into the substrate.

Convolute – 1. Refers to the arrangement of floral or foliar organs in a bud when each organ or segment has one edge overlapping the adjacent organ or segment: a form of imbricate arrangement. See *contort.* 2. (*of leaves*) vernation in which one leaf is rolled up inside another. 3. vernation of two leaves at a node, in which one half of each leaf is exposed, and the other half is wrapped inside the other leaf.

Copse – a dense growth of trees, shrubs, or bushes.

Corcle – a plant embryo, plumule, or plumule plus radicle.

Cordate – heart-shaped, with the notch lowermost of the base of a leaf, like the notched part of a heart. Contrast *obcordate.*

Coriaceous – leathery; stiff and tough, but flexible. Compare *corneous.*

Corm – short, thick underground storage shoot with the prevalence of stem tissues. A fleshy, swollen stem base, usually underground and functioning in the storage of food reserves, with buds naked or covered by very thin scales; a type of rootstock.

Corneous – horny in texture; stiff and hard, but somewhat tough. Compare *coriaceous.*

Corolla – a collective term for the petals of a flower. Compare *calyx.*

Corona – 1. (*in flowering plants*) a ring of structures that may be united in a tube, arising from the corolla or perianth of a flower and standing between the perianth lobes and the stamens. The trumpet of a daffodil is a corona. 2. (*in grasses*) a hardened tissue ring surmounting some species' lemma.

Cortex (pl. *cortexes or cortices*) – the external layer of the primary stem or root. Cells that make up the primary tissue of the root and stem. In lichens, the skin or outer layer of thallus tissue covers the medulla. Fruticose lichens have one cortex encircling the branches, even flattened, leaf-like forms; foliose lichens have different upper and lower cortices; crustose, placodioid, and squamulose lichens have an upper cortex, but no lower cortex; and leprose lichens lack any cortex.

Corticolous – growing on bark or wood with the bark stripped off. Compare *lignicolous.*

Corymb (adj. *corymbose*) – an inflorescence with branches arising at different points but reaching about the same height, giving the flower cluster a flat-topped appearance.

Costa – a rib.

Any duplication (copies, uploads, PDFs) is illegal.

Costapalmate – having a definite costa (midrib), unlike the typical palmate or fan leaf, but with the leaflets arranged radially as in a palmate leaf.

Cotyledon (or *seed leaf*) – embryonic leaf. The first leaf that appears on a seedling. The primary leaf or leaves of a plant embryo develop into the seed leaf or the first set of leaves upon germination.

Craspedodromous – pinnate venation with secondary veins terminating at the margins, often as teeth.

Crateriform – in the shape of a saucer or shallow cup; hemispherical or more shallow.

Cremnophyte – a plant adapted to growing on, especially hanging from, cliff faces or crevices. Compare *chasmophyte*.

Crenate – having blunt or rounded teeth; scalloped.

Crenulate – minutely scalloped.

Crisped – finely curled, as with the edges of leaves and petals.

Cristarque cell – a sclereid containing a druse and has the lignin deposited exocentrically on the cell wall to form a cup shape, or in cross-section, a U-shape.

Cross – to make something interbreed, the act of hybridization.

Cross-pollination – between genetically different plants.

Crown – consists of leaves and branches. See *canopy*.

Cruciform – cross-shaped.

Crustaceous – hard, thin, and brittle.

Crustose – forming a closely applied surface layer or crust.

Cryptogam – the "lower plants" that produce spores and do not have stamens, ovaries, or seeds; literally, plants whose sexual reproductive organs are not conspicuous. This group typically includes ferns, bryophytes, algae, and sometimes fungi (including lichenized fungi). Compare *phanerogam*.

Cucullate – hood-like or hooded, commonly referring to the shape of leaves or petals (e.g., Pelargonium cucullatum). Similarly derived terms include *cuculliform* and *cuccularis*.

Culm – in grasses, sedges, rushes, and some other monocotyledons, an aerial stem bearing the inflorescence, extending strictly from the base of the plant to the lowest involucral bract (or base of the inflorescence).

Cuticle – a relatively impermeable surface layer on the epidermis of leaves and fruits.

Cultigen – a plant whose origin or selection is primarily due to intentional human activity.

Cultivar – a term derived from "*cultivated variety*" denoting an assemblage of cultivated plants distinguished by characteristics (e.g., morphological, physiological, cytological). When reproduced (sexually or asexually), the assemblage retains its distinguishing characteristics. A cultivar may arise in

 Copyright © Sterling Education

cultivation or naturally as a variant that is of horticultural interest. Cultivar names are written with single quotation marks around them (e.g., 'Blue Carpet' or 'Alba'). Names established after January 1, 1959, must be in a common language (i.e., not in Latin), but names in Latin before this date are retained.

Cultivar epithet – the defining part of a name that denominates a cultivar. Cultivars are designated by fancy (q.v.) epithets appended either to the scientific name or to the common name of the taxon to which they belong; they are not italicized but placed in single quotation marks (e.g., *Rubus nitidoides* 'Merton Early'. 'Merton Early' is the cultivar epithet).

Cuneate – wedge-shaped, with straight sides converging at a base.

Cupule – a cup-shaped structure composed of coalescent bracts, such as the cup of an acorn. See *calybium.*

Cupular – shaped like a cupule.

Cupulate – bearing cupules.

Cupuliform – nearly hemispherical, shaped like a cupola or dome.

Cusp – a hard, pointed tip, stiffer and more formidable than a mucro, hence cuspidate.

Cuspidate – tipped with a cusp, as with some leaves.

Cuticle – a waterproofing layer covering the epidermis of aerial plant surfaces and composed of polymers cutin or waxes.

Cutting – an apical tip of shoot structure, root, or leaf cut from a plant and used for asexual vegetative propagation.

Cyathium (pl. *cyathia*) – an inflorescence of unisexual flowers surrounded by involucral bracts, especially the flowers of *Euphorbia.*

Cyathophyll – in *Euphorbia*, the bract-like structure on which the involucre usually sits but does not always occur in twos. They may be brightly colored and confused with petals.

Cylindrical – rod-like and two to three times as long as wide. Compare *baculiform.*

Cynaroid – see *carduoid.*

Cyme (adj. *cymose*) – inflorescence in which the main axis and all lateral branches end in a flower (each lateral may be repeatedly branched).

Cymose – having a cyme or cymes.

Cypsela – a dry, one-seeded, indehiscent fruit formed from an inferior ovary.

Any duplication (copies, uploads, PDFs) is illegal.

D

Deciduous [Latin, *hanging down* or *falling*] – shedding foliage at the end of the growing season. Unlike *conifers*, *deciduous* trees drop their leaves in the fall. Generically known as *hardwoods*, *deciduous* species include oak and maple. Dehiscing and falling seasonally, as with bark, leaves, or petals. Contrast *persistent.*

Declinate – curving downward and then upward at the tip. Often qualified, e.g., declinate-ascendant.

Decompound – divided into more than one level (e.g., bipinnate leaves), in which the leaflets of what would otherwise be a pinnate leaf are divided.

Decorticate – 1. to shed the outer bark of a tree, usually seasonally, as part of the natural growth cycle. 2. to strip the peel, crust, bark, or other surface tissues from a plant or harvested material, such as extracting fiber from harvested *Agave* leaves.

Decumbent – having branches growing horizontally along the ground but turned up at the ends.

Decurrent – extending downward beyond the insertion point (e.g., when the base of a leaf or a fungal gill is prolonged downward along the stem in a raised line or narrow wing).

Decussate – opposite with successive pairs borne at right angles to the last; generally applied to the arrangement of leaves.

Definite – of a constant number (e.g., twice as many stamens as petals or sepals (or less), or an inflorescence ending in a flower or an aborted floral bud, typically a cymose inflorescence. Contrast *indefinite.*

Deflexed – bent downward. Contrast *inflexed.*

Dehiscent – fruits that open. Breaking open at maturity to release contents refers to opening fruits to release seeds, anthers to release pollen, and sporangia to release spores. Contrast *indehiscent.*

Deltoid – shaped like the uppercase Greek letter **Δ** (i.e., like an equilateral triangle).

Dendroid – tree-like; branching like a tree.

Dentate – toothed, especially in reference to leaf margins.

Denticulate – finely toothed; a diminutive form of dentate.

Deserticolous – inhabiting a desert.

Determinate – limited, usually in growth. Contrast *indeterminate.*

Diaspore – any reproductive part of a plant adapted for dispersal and for establishing new plants; a seed or other parts (e.g., specialized buds, branches, inflorescences, or fruits).

Dichasium – a cymose inflorescence with all branches below the terminal flower in regular opposite pairs. Compare *monochasium* and *pleiochasium.*

 Copyright © Sterling Education

Dichlamydeous – having a perianth, which is divided into a separate calyx and corolla. Compare *homochlamydeous*.

Dichotomous – forking into two equal branches. This may result from an equal division of the growing tip or sympodial, in which the growing tip is aborted and replaced. Typically refers to the mode of branch growth (e.g., Aloidendron dichotomum) but also to other organs, such as the venation patterns on leaves, the thorns of various species of Carissa (which morphologically are branches), and the thalli or hyphae of various algae and fungi.

Dichotomous branching – when the terminal bud always divides in two.

Dicot – having two seed leaves.

Dicotyledon (or *dicot*) – a flowering plant whose embryo has two or more cotyledons (seed leaves). Contrast *monocotyledon*.

Dioecious (*of vascular plants*) – having male and female reproductive structures that develop only on different individuals and never on the same individual. Contrast *monoecious*.

Dioicous (*of a bryophyte gametophyte*) – having male and female reproductive structures that develop only on different individuals and never on the same individual. Contrast *monoicous*.

Dioecious plants – unisexual flowers with one sex on each plant; in effect, male and female plants

Digitate – with segments spreading from a common center (e.g., fingers of a hand). See *palmate* and *palmatisect*.

Dimorphic – occurring in two forms (concerning shape or size). For example, stamens, fronds, or leaves. See *monomorphic* (i.e., having a single form) and *polymorphic* (i.e., having many forms).

Diploid (2N) – having two complete sets of chromosomes in the nucleus of a sporophyte cell (i.e., one set from each of the parental gametes). This is often expressed as $2N$, where N = the number of chromosomes in the haploid (1N) gamete.

Diplostemonous – having stamens arranged in two whorls, with the outer whorl alternating with the petals while the inner whorl is opposite the petals. Compare *obdiplostemonous* and *haplostemonous*.

Disc (or *disk*) – a plate or ring of structures derived from the receptacle and occurring between whorls of floral parts. In some groups, especially Sapindales, the nectary is in the form of a prominent disk. In daisies, the central part of the capitulum is a disk; hence, flowers are borne by *disk flowers* or *florets*.

Discoid – 1. resembling a disc or plate, having both thickness and parallel faces and a rounded margin. 2. describes the flower head of *Asteraceae* where there are no ray florets but only disc florets.

Discolorous (*of leaves*) – having upper and lower surfaces of different colors.

Disjunct – occurring in widely separated geographic areas, distinctly separate; applies to a discontinuous range in which one or more populations are separated from other potentially interbreeding populations with sufficient distance to preclude gene flow between them.

Any duplication (copies, uploads, PDFs) is illegal.

Disk floret – occurring typically in the disk of the capitulum of flowers in the family *Asteraceae* and, to some extent, in other plants that bear a flowering head with a disk, such as Scabiosa.

Dispersal – the method by which plants spread their seeds (animal, expulsion, water, wind).

Dissected – deeply divided; cut into many segments.

Dissepiment – a partition or septum in a plant part, usually referring to septa between the loculi of capsules or other fruits with multiple partitions.

Distal – remote from the point of origin or attachment; the free end. Contrast *proximal.*

Distichous – arranged in two opposite rows (and hence in the same plane).

Distinct – separate or free; not united.

Distyly – the condition in which the flowers of a species occur in two forms that differ only by the length of the style and stamens, and flowers of only one of these forms appear on any one plant. Compare *heterostyly.*

Diurnal – of the day; occurring or opening in the daytime.

Divaricate – wide-spreading.

Divergent – spreading in different directions, generally upward.

Division – a taxonomic rank below kingdom in the standard taxonomic hierarchy. "*Division*" is generally used only for plants and is the approximate botanical equivalent of the term *phylum*, which is used for animals and other kingdoms.

Domatia (sing. *domatium*) – any hollow structure formed by a plant inhabited by animals such as ants or mites.

Dorsal [Latin *dorsum* – a ridge or the back of an animal]. Partly because the term originally referred to animals rather than plants, usage in botany is arbitrary according to context and source.
In general, dorsal refers to "*the rear or back or upper surface*," but in botanical usage, such concepts are not always clearly defined and may be contradictory. For example, facing away from the axis (abaxial) in a lateral organ of an erect plant, facing away from the substrate in any part of an erect plant. For example, the upper surface of a horizontal leaf (adaxial) or the upper part of the crown of the plant facing away from the substrate in a prostrate or climbing plant or floating leaves such as those of Nymphaea. Related anatomical terms of location include *ventral* and *lateral.*

Dorsifixed – attached at or by the back (e.g., anthers on a filament).

Dorsiventral – having structurally and visibly different upper and lower surfaces (e.g., some leaves). Compare *bilateral* and *isobilateral.*

Double fertilization – when two brother male gametes fertilize two female sister cells.

Drip tip – a long, narrow, acuminate, caudate, or cuspidate extension at the tip of a leaf or leaflet. Commonly an adaptation to rainy conditions, as it promotes the shedding of water by its dripping from

 Copyright © Sterling Education

the narrow tip. A drip tip is not anatomically descriptive as an acuminate or cuspidate but a description of the functional shape that aids dripping, irrespective of the specific geometry of the shape itself.

Drupe (or *kernel*) – a succulent fruit formed from one carpel; the single seed is enclosed by a stony layer of the fruit wall (e.g., in peaches and olives).

Drupelet – a small drupe formed from one carpel in an apocarpous flower. Drupelets usually form a compound fruit, as in *Rubus*, but they may become widely separated, as in *Ochna*.

Druse – a globular mass of calcium oxalate crystals, usually with the crystals radiating from an organic core.

E

Ebracteate – lacking bracts; synonymous with *ebracteolate*.

Ecological amplitude – the range of environmental conditions in which an organism can survive.

Edaphic – of or influenced by the soil.

Elaiosome – an external structure attached to the seed of many species of plants. Elaiosomes generally look fleshy; some species are rich in oils or other nutritious materials. Their functions vary and are not always obvious; commonly, they attract ants or other animals that aid in dispersal, but they may repel other animals from eating the seed.

Electron Transport Chain (ETC) – a series of chemical reactions that produce 34 ATP molecules and H_2O from the carrier molecules produced in the Krebs cycle.

Elephophily – pollination with pollen or spores distributed by the feet of elephants.

Ellipsoid – a three-dimensional elliptical shape in all sections through the long axis.

Elliptical (or *elliptic*) – planar, shaped like a flattened circle, symmetrical about the long and short axes, tapering equally to the tip and base; oval.

Elongation zone root – a region of expanding cells.

Emarginate – notched at the apex (notch usually broad and shallow).

Embryo – female gametophyte of flowering plants; contained by a seed prior to germination.

Emergent – a plant taller than the surrounding vegetation or, among aquatic plant species, one that bears flowers and commonly leaves above the water's surface. Aquatic examples include water lilies, reeds, and papyrus. Some pondweeds, such as *Stuckenia*, do not emerge until they flower, at which time only their flowers appear above the water surface.

Enantiostyly – the condition in which the gynoecium protrudes laterally, to the right (dextrostyly) or the left (sinistrostyly) of the androecium (e.g., Senna).

Any duplication (copies, uploads, PDFs) is illegal.

Endemic – having a natural distribution restricted to a particular geographic region. Compare *native*.

Endergonic chemical reaction – free energy (ΔG) is required to proceed.

Endocarp – the innermost layer of the wall of a fruit; in a drupe, the stony layer surrounds the seed.

Endodermis – the innermost layer of the cortex of vascular plant roots present in the stems of pteridophytes. The radial walls are impregnated with suberin to form a permeability barrier known as the Casparian strip.

Endophytic fungi – grows inside the plant body.

Endophloic (or *endophloeodal*) (*of crustose lichens*) – having the thallus growing within rather than upon the bark of trees. Compare *epiphloedal* and *corticolous* (i.e., growing on the surface of wood or bark) and *endolithic* (i.e., growing within a rock).

Endosperm – 1. (*angiosperms*) a nutritive tissue surrounding the embryo of the seed, usually triploid, originating from the fusion of both polar nuclei with one gamete after the fertilization of the egg. 2. (*gymnosperms*) the prothallus within the embryo sac.

Endosperm – triploid (sometimes diploid) nutrition tissue originated from second fertilization.

Endosperm haploid – nutrition tissue originated from the female gametophyte.

Endospory – the production of spores that germinate into a reduced multicellular gametophyte within the spore wall. Contrast *exospory*.

Energy – the ability to do work or cause change.

Ensiform – shaped like the blade of a sword.

Entire – 1. not divided. 2. (*of a margin*) smooth and not lobed (though possibly wavy or scalloped).

Entomophily – pollination whereby insects distribute pollen or spores.

Ephemeral – short-lived. See *caducous*.

Epicalyx – an involucre resembling an outer calyx (e.g., as in *Hibiscus*).

Epicarp – the outer layer of the wall of a fruit (i.e., the *skin*).

Epicormic – refers to buds, shoots, or flowers developing from the old wood of trees, especially after injury or fire.

Epicotyl – first internode of the stem. The part of the plant axis or stem between the cotyledonary node and the first foliage leaves.

Epicuticular wax – a layer of crystalline or amorphous wax deposited on the surface of the cuticle.

Epidermis – complex surface tissue. *The outermost layer of plant cells*. An organ's outermost layer of cells usually only one cell thick.

 Copyright © Sterling Education

Epigynous – borne on the ovary; describes floral parts when attached above the level of the ovary and arising from tissue fused to the ovary wall. Compare *hypogynous* and *perigynous*.

Epilithic – growing on stone. Compare *lithophytic*, a plant growing on stone.

Epinecral – dead (or *necral*) tissue above the surface of the cortex of a lichen.

Epipetalous – of stamens that are attached to the petals.

Epipetric – growing on rock or stone, lithophytic, epilithic.

Epiphloedal – growing on the surface of the bark. Contrast *endophloedal* (i.e., growing inside, not on, the bark) and *epilithic* (i.e., growing on rock, not bark).

Epiphyte – a plant, alga, or fungus that grows on another plant without deriving nourishment from it but using it for support.

Epiphytic –an epiphyte living on the surface of a plant. Compare *epilithic* and *lithophytic*.

Epitepalous – of stamens that are attached to the tepals.

Epithet – the adjectival component in a binomial scientific name, usually more precisely called a specific epithet; the final word or combination of words in the name of more than one word (other than a term denoting rank) that denominates an individual taxon. The simplest and most common example is the second word in a two-word name of a species, such as "*mirabilis*" in *Welwitschia mirabilis*.

Epizoochory – seed dispersal when seeds or fruits adhere to the outside of vertebrate animal bodies.

Epruinose – not pruinose.

Equitant (*of a leaf*) – folded lengthwise and clasping another leaf.

Erect – upright, perpendicular to the ground or point of attachment. Compare *patent* (spreading) and *erecto-patent* (between erect and patent).

Ericoid – having leaves like those of the European heaths (*Erica*); small and sharply pointed.

Erose (*of a margin*) – irregular as though nibbled or worn away.

Eustele – vascular bundles in a ring.

Even-pinnate – having an even number of leaflets in a compound leaf; synonymous with *paripinnate*.

Evergreen – not deciduous; it has leaves all year.

Ex – in nomenclature, indicating that the preceding author proposed the name but did not legitimately publish it and that the succeeding author referred to the first author when legitimately publishing it.

Exalbuminous – in seeds of a given species, having no endosperm (i.e., no albumen); for example, in *Fabaceae* and *Combretaceae*.

Exergonic chemical reaction – free energy (ΔG) is released.

Any duplication (copies, uploads, PDFs) is illegal.

Exocarp – the outer layer of the pericarp, often the skin of fleshy fruits.

Exodermis – the outermost layer of cortex.

Exospory – the production of spores that germinate into free-living multicellular gametophytes. Contrast *endospory.*

Exotestal – the outer layer of the testa (seed coat). It is derived from the outer integument of the ovule.

Exotic – not native; introduced from another region or country.

Exserted – projected beyond (e.g., stamens beyond the corolla tube).

Exstipulate – lacking stipules.

Extrastaminal – outside the stamens or androecium, usually referring to the location of a nectary disk.

Extrorse (*of anther locules*) – opening toward the outside of the flower. Contrast *introrse* and *latrorse.*

F

F1 hybrid – a single cross; a plant breeding term for the result of a repeatable cross between two purebred lines.

F2 hybrid – a plant breeding term for the result of a plant arising from a cross between two F1 hybrids; it may refer to self-pollination in a population of F1 hybrids.

Fabiform – shaped like a kidney bean.

Facultative – of parasites, optional. Compare *obligate.*

Falcate – curved like the blade of a *scythe.*

Family – a taxonomic group of one or more genera with features or ancestry in common; the term for the principal rank between order and genus.

Farina – powdery, pale yellow crystalline secretion of flavonoids in *Primula* and other species.

Fascicle (adj. *fasciculate*) – a cluster (e.g., a tuft of leaves arising from the same node).

Fasciculate – branching in clusters like a bundle of sticks or needles; having *fascicles.*

Faveolate – honeycombed; having regular, angled pits. Compare *foveolate.*

Faucal – pertaining to the *fauces*; located in the throat of a *calyx* or *corolla.*

Fauces – the throat of a *calyx* or *corolla*; the conspicuously widened portion between the mouth and the apex of the tube. In *Boraginaceae*, the site of distinctive appendages.

 Copyright © Sterling Education

Felted – covered with very dense, interlocked, and matted hairs with the appearance or texture of felt or woolen cloth.

Female flower – see *pistillate flower*.

Fenestrate – having translucent areas. See *perforate* (i.e., with holes).

Fermentation – if sufficient oxygen is not present in the cell, glycolysis is followed by a different pathway that produces little energy and either alcohol or lactic acid, depending on the organism.

Ferruginous – ruddy or rust-colored.

Fertile – capable of producing fruit; of flowers when they produce seed or of anthers containing pollen.

Fertilization – the union of male and female gametes during sexual reproduction.

Fiber – 1. a fiber cell. 2. any flexible, strong, stringy, and very elongated structure.

Fibers – long and narrow sclerenchyma cells.

Fiber cell – a type of cell found in sclerenchyma, it is much elongated and dies soon after an extensive modification of its cell wall. The cell wall is usually thickly lignified but is sometimes gelatinous.

Fibrous root system – no primary root visible.

Fiddleheads – spiral tops of young fern leaves.

Filament – 1. the stalk of a stamen. 2. any very narrow, thread-like structure one or a few cells thick.

Filamentous – consisting of filaments or fibers, hairlike.

Filiform – thread-like (e.g., stamen filaments or leaf shapes).

Fimbria (pl. *fimbriae*) – slender hair-like process.

Fimbriate – fringed.

Fissure – a split or crack, often referring to fissured bark; a line or dehiscence opening.

Fistule – a tube-shaped cavity.

Fistulose – hollow; usually applied to a tube-shaped cavity, as in a reed.

Flabellate – fan-shaped.

Flaccid – limp; tending to wilt. Compare *turgid*.

Flexistyly – depending on the degree of maturation of the stamens, the style moves up or down.

Flexuose – bent alternately in different directions; zigzag.

Floccose – has a soft and woolly covering of hair.

Any duplication (copies, uploads, PDFs) is illegal.

Flora – 1. all the plants growing in a specific region or country. 2. an enumeration of them, generally with a guide to their identification.

Floral diagram – a graphical representation of flower structure, usually a schematic cross-section of a young flower.

Floral formula – a description of flower structure using numbers, letters, and various symbols.

Floral envelope – see *perianth.*

Floral leaves – the upper leaves at the base of the flowering branches.

Floral tube – an imprecise term used as a synonym of hypanthium, corolla tube, or calyx tube.

Floral units (FU) – elements of a generative system, such as *fructifications.*

Floret – a small flower, usually referring to the individual true flowers clustered within an inflorescence, particularly those of the grasses and the pseudanthia of daisies.

Flower – the sexual reproductive structure of the angiosperms, typically with a *gynoecium, androecium, perianth,* and an *axis.* Compact generative shoot with sterile male and female zones, in that order. Floral unit with sterile male and female zones. ***Actinomorphic flower*** – having multiple planes of symmetry, formula. ***Zygomorphic flower*** – having only one plane of symmetry. ***Perfect flower*** – having both sexes male / female flower.

Foliate – preceded by a number, which means having a certain number of leaflets; for example, 3-foliate means "having three leaflets."

Foliaceous – a growth habit of certain lichens, algae, and fungi that prefer to grow on the leaves of vascular plants.

Follicle – a dry fruit formed from one carpel, splitting along a single suture to which the seeds are attached. Compare *pod (of legume).*

Foliole – a small leaf-like appendage on the front or back.

Foliose – leaf-like; flattened like a leaf.

Forb – any non-woody flowering plant that is not a grass, sedge, or rush.

Forest – a large, densely wooded area filled with trees and plants. Vegetation dominated by trees with single trunks (including closely arranged trees with or without an understory of shrubs and herbs).

Forma *(or form)* – a taxonomic category subordinate to species and within the taxonomic hierarchy, below varieties, and usually differentiated by a minor character.

Foveolate – having regular tiny pits. Compare *faveolate.*

Free – not united with other organs of the same type; not attached at one end.

Free central *(of placentation)* ovules attached to a free-standing column in the center of a unilocular ovary.

 Copyright © Sterling Education

Frond – a leaf of a fern, cycad, or palm.

Frutescent – shrub-like (*fruticose*) or becoming shrub-like.

Fruticose – shrubby; having the branching character of a shrub.

Fruit – ripe floral unit (FU). A seed-bearing structure present in all angiosperms, formed from the mature ovary and sometimes associated floral parts upon fertilization.

Fugacious – disappearing, falling off, or withering. Compare *persistent* and *caducous*.

Funicle (or *funiculus*) – the stalk of an ovule.

Funnelform – having a form gradually widening from the base to the apex; funnel-shaped.

Furcate – forked, usually applied to a terminal division; with two long lobes.

Fused – joined together.

Fusiform – initials cambium cells, which make vessel elements. Rod-shaped and narrowing gradually from the middle toward each end; spindle-shaped.

G

Galbulus (*in gymnosperms*) – a fleshy cone (i.e., *megastrobilus*) chiefly relates to cones borne by junipers and cypresses, often mistakenly called berries.

Gall – abnormal swelling of plant tissue. A gall is a lump on the side of a tree trunk or branch caused by an insect laying its eggs inside the bark. Because of their dark discoloration, the galls from oak trees have been used to make ink and dye for centuries. *Various parasites, from viruses, fungi, and bacteria to other plants, insects, and mites, cause abnormal outgrowth on external plant tissues.*

Gamete – a cell or nucleus that fuses with another of the opposite sex during sexual reproduction.

Gametophore – specialized structures on the gametophytes of some bryophyte species. For example, many species in the order Marchantiales; in such species, the gametes are produced on the gametophores, which amount to sex organs.

Gametophyte – the haploid (1N) multicellular phase in alternating generations of plants and algae that bear gametes. In bryophytes, the gametophyte is the dominant vegetative phase; in ferns and their allies, it is a small free-living plant known as the prothallus; in gymnosperms and angiosperms, the gametophytes are reduced to microscopic structures dependent on the sporophyte, male gametophytes contained in pollen grains and females contained within the ovules.

Gemma – an asexual reproductive structure found in liverworts and mosses.

Gene pool – the complete range of genetic variation found within a population.

Any duplication (copies, uploads, PDFs) is illegal.

Generative shoot system – all generative shoots together.

Germinate (or *sprout*) – produces buds or branches, a process of growth or development.

Genus (pl. *genera*) – a group of one or more species with features or ancestry (or both) in common. The genus is the principal category of taxa intermediate in rank between family and species in the standard nomenclatural hierarchy.

Generic name – the name of a taxonomic genus, such as *Acacia* and *Eucalyptus*.

Genotype – the genetic makeup of an organism.

Geophilous – growing or rooting in the ground.

Germination – 1. (*of seeds*) the complex sequence of physiological and structural changes that occur from resting to growth stage. 2. (*of a pollen grain*) production of a pollen tube when contacting a receptive stigma. 3. (*of a spore of fungi or bacterium*) change of state (e.g., from resting to vegetative).

Gibbous (or *gibbose*) (*of part of an organ*) – swollen, usually with a pouch-like enlargement at the base.

Glabrescent – becoming glabrous, almost glabrous.

Glabrous – 1. lacking surface ornamentation such as hairs, scales, or bristles; smooth.
2. (*in lichenology*) having no indumentum.

Gland – a secretory structure within or on the surface of a plant.

Glandular hair – hair tipped with a gland.

Glaucous – describes the external surface of a plant part with a whitish covering, sometimes with a blueish cast. Often applied to plants with a woolly or arachnoid surface, but properly referring to pruinose surfaces, meaning those with a waxy bloom. For example, the surface of the young leaves of many *eucalypts* and *xerophytes*.

Globose (or *globular*) – roughly spherical. See *subglobose*.

Globulose – approximately spherical.

Glochid – a tiny, barbed hair or bristle (e.g., the fine defensive hairs in cactus species such as *Opuntia*).

Glucose ($C_6H_{12}O_6$) – a six-carbon sugar produced by photosynthesis.

Glumes – bracts subtending the floret(s) of a sedge or similar plant; in grasses forming the lowermost organs of a spikelet (there are usually 2, but 1 is sometimes reduced; or rarely, both are absent).

Glutinous – sticky.

Glycolysis – a process in which one glucose molecule is broken in half by enzymes in the cytoplasm, producing 2 molecules of pyruvic acid and only 2 molecules of ATP.

Graft –1. artificial union of plant parts. 2. a plant shoot suitable for grafting; a scion, sucker, or branch.

Copyright © Sterling Education

Graft chimera (or *graft hybrid)* – a taxon whose members consist of tissue from two or more plants in intimate association originated by grafting. The nomenclature of graft hybrids is governed by the International Code of Nomenclature for Cultivated Plants.

Grana – neatly folded layers of thylakoids resembling stacks of pancakes.

Granular (*of a surface*) – covered with small, rounded protuberances.

Grass – a plant of the family *Poaceae.*

Grassland – low vegetation dominated by grasses.

Groundcover – 1. dense vegetation that covers the ground. 2. describes a plant that covers the soil surface so densely that it smothers all beneath it.

Ground meristem – primary meristem, which makes cortex and pith.

Ground tissue – the same as *parenchyma* but only applied to tissue.

Group – a formal category equivalent to or below the rank of the genus which distinguishes 1) an assemblage of two or more cultivars within a species or hybrid, 2) plants derived from a hybrid in which one or more of the parent species is not known or is of uncertain origin, and 3) a range of cultivated plants of a species or hybrid which may exhibit variation but share one or more characters, which makes it worth distinguishing them as a unit.

Guard cells – epidermal cells that open and close to let water, oxygen, and carbon dioxide pass through the stomata. Each of the two cells surrounding the stoma controls gas exchange between the apoplast of the plant and the external environment.

Guttate – having droplet-shaped spots. Compare *punctate* and *maculate.*

Guttation – the secretion of liquid water from uninjured plant parts. See *hydathode.*

Guttulate – having or appearing to be spotted with oil droplets; of spores, having oil droplets inside.

Gymnosperm – a seed-bearing plant with unenclosed ovules borne on the surface of a sporophyll. Gymnosperms are among the oldest clades of vascular plants and represented by 1,000 extant species worldwide, including, among others, *conifers*, *Ginkgo*, *Gnetum,* and *cycads*. Compare *angiosperm.*

Gynobasic – a style arising near the base of the gynoecium (e.g., between the ovary lobes).

Gynodioecious – of a species, with some plants bearing only bisexual flowers and others bearing only female flowers.

Gynoecium – collective term for pistils/carpels. Gynoecium can comprise: 1. a single carpel means simple pistil; this is *monomery*. 2. two or more fused carpels (i.e., compound pistils) are *syncarps*. 3. Two or more unfused carpels (i.e., two or more simple pistils) are *apocarpy. A variant with several compound pistils does not exist in nature. Carpel number in a compound pistil; locules, points of placentation, and the number of styles, stigma, and ovary lobes*. Contrast *androecium.*

Gynomonoecious – of a species with bisexual flowers and female flowers on the same plant.

Any duplication (copies, uploads, PDFs) is illegal.

Gynophore – a stalk supporting the gynoecium and situated above the level of insertion of the other floral parts.

Gynostegium – a compound organ in milkweeds (*Asclepiadaceae*) and orchids formed by the fusion of the filaments of the stamens with the style. Also known as the *column*.

H

Habit – the general external appearance of a plant, including size, shape, texture, and orientation.

Habitat – where a plant lives, the environmental conditions of its home.

Hafter – a flat attachment forms when the thallus of a fruticose or foliose lichens encounters a substrate.

Hair – a single elongated cell or row of cells borne on the surface of an organ.

Half-inferior ovary – partly below and partly above the level of attachment of the other floral parts. Compare *inferior ovary* and *superior ovary*.

Halonate – having a transparent coating or being of a spore's outer layer.

Halophyte – plants adapted to the over-presence of NaCl. A plant adapted to living in highly saline habitats, a plant that accumulates high concentrations of salt in its tissues.

Hand-pollination – the controlled act of pollination that excludes the possibility of open-pollination.

Haploid (1N) – having one set of chromosomes. The complement of chromosomes in each of the cells of the gametophyte (i.e., *nucleus of a gamete* and *spore*). Expressed symbolically as N, where N = the gametic number of chromosomes. Compare *diploid, triploid,* and *tetraploid.*

Haplostemonous – having a single series of stamens equal to the proper number of petals and alternating with them. Compare *diplostemonous*, and *obdiplostemonous*.

Hapter – an attachment that may form when a foliose lichen thallus encounters a substrate.

Harmomegathy – the process by which pollen grains in arid environments close off their apertures to avoid losing water.

Hastate – triangular in outline,with the basal lobes pointing outward so that the base appears truncate; it may refer only to the base of a leaf with such lobes. Compare *sagittate,* which refers to basal lobes pointing backward.

Haustoria – sucker roots of parasitic plants.

Haustorium – in parasitic plants, a structure developed for penetrating the host's tissues.

Head – see *capitulum*, a pseudanthium.

 Copyright © Sterling Education

Heathland – vegetation dominated by small shrubs, which usually have ericoid leaves.

Heartwood – a non-functional part of wood. The dense inner center part of the tree gives it strength.

Helicoid – coiled; of a cymose inflorescence, when the branching is repeatedly on the same side (the apex is often recurved). Compare *scorpioid.*

Heliophytes – plants adapted to full sun.

Hemi-legume – legume fruit in which the seed or seeds and one pod valve are dispersed as a unit. The valve catches the wind and blows away with the seeds, as in *Acacia tenuifolia* and *Peltogyne paniculata.*

Hemiparasites – photosynthetic plants that feed partly on other plants.

Herb – any vascular plant that does not develop a woody stem at any point during its life cycle (e.g., a violet).

Herbaceous – not woody; usually green and soft in texture.

Herbarium (pl. *herbaria*) – a collection of preserved, usually pressed and dried, plant material used for identification and comparison; also, a building in which such collections are stored.

Hermaphrodite – a synonym of bisexual.

Heteroblastic – having parts, especially leaves, that are distinctly different between the juvenile and adult stages.

Heteromorphic – having two or more distinct morphologies (e.g., size and shape).

Heterophylly – when one plant has more than one leaf type.

Heterosporic – with male and female spores.

Heterospory – the production of spores of two sizes (small and large) by the sporophytes of land plants.

Heterostyly – the condition of a species having flowers with distinctive style and stamen lengths, but with all the flowers of any one plant being identical. See *distyly.*

Heterotrophs (or *consumers*) – organisms that cannot make food.

Hilum – the scar on a seed coat where it separates from its stalk (funicle).

Hip – the fruit of a rose plant.

Hippocrepiform – horseshoe-shaped.

Hirsute – bearing coarse, rough, longish hairs. See indumentum.

Hispid – bearing long, erect, rigid hairs or bristles, harsh to touch.

Hoary – covered with a greyish-to-whitish layer of very short, closely interwoven hairs, giving a frosted appearance.

Any duplication (copies, uploads, PDFs) is illegal.

Holotype – a type chosen by the author of a name. Compare *lectotype.*

Homochlamydeous – with a perianth not divided into a separate calyx and corolla. Contrast *dichlamydeous.*

Homoiohydric – plants that save water.

hort. – (*never capitalized*) *Of gardens, an author citation used in two ways:* 1. as a name misapplied by gardeners. 2. as an invalid name derived from horticultural writings of confused authorship.

Hyaline – translucent; usually delicately membranous and colorless.

Hybrid – a plant produced by the crossing of parents belonging to two different named groups (e.g., genera, species, varieties, subspecies, forma (i.e., the progeny resulting within and between two plants). An F1 hybrid is the primary product of such a cross. An F2 hybrid is a plant arising from a cross between two F1 hybrids (or from the self-pollination of an F1 hybrid).

Hybrid formula – parental names of a hybrid joined by a multiplication sign (e.g., *Cytisus ardonoi* × *C. purgans*).

Hydrophily – pollination whereby the flow of water distributes pollen.

Hydrophytes – plants growing in water and frequently using water for support.

Hygrophytes – terrestrial or partly submerged plants adapted to excess water.

Hypanthium – a tube or cup-like structure in a flower that includes the bases of sepals, petals, and stamens, and may or may not be connected to the ovary.

Hyper-resupinate (*botany*) – describing leaves or flowers that are in the usual position but are borne on a petiole or pedicel, respectively that is twisted 360 degrees. The term describes organs, such as orchid flowers, which are usually resupinate. Compare *resupinate.*

Hypocarpium – an enlarged fleshy structure that forms below the fruit from the receptacle or hypanthium.

Hypocotyl – of an embryo or seedling, the part of the plant axis below the cotyledon and node but above the root. It marks the transition from root to stem development.

Hypocrateriform – see *salverform.*

Hypogynous – borne below the ovary; used to describe floral parts inserted below the ovary's level of insertion. Compare *epigynous*, and *perigynous.*

Hypothallus – the outgrowth of hyphae from under the margin of the thallus of a crustose lichen, connecting the island-like areoles into a single lichen.

Hysteranthous (or *histeranthous*) – growth in which new leaves appear after flowering. Compare *proteranthous* and *synanthous.*

Copyright © Sterling Education

I

Idioblast – a cell, especially of a leaf, differs markedly from surrounding cells and often synthesizes specialized products such as crystals.

Illegitimate name – not abiding by the rules of the botanical Codes.

Imbricate – overlapping each other of perianth parts, edges overlapping in the bud (a convoluted arrangement is a particular form of imbrication). Dormant buds of many deciduous species are imbricately covered with protective cataphylls called bud scales.

Imparipinnate – pinnate leaf with an odd number of pinnae (terminated by a single leaflet). Compare *paripinnate*.

Inbreeding – the production of offspring between closely related parents leading to a high degree of similarity; self-fertilization is the most intense form of inbreeding.

Incertae sedis – of unknown taxonomic affinity; relationships obscure.

Incised – cut deeply and (usually) unevenly (a condition intermediate between toothed and lobed).

Included – enclosed, not protruding (e.g., stamens within the corolla).

Incomplete flower – one or more of its usual parts, such as carpels, sepals, petals, pistils, or stamens.

Incurved – bent or curved inward; of leaf margins, when curved toward the adaxial side.

Indefinite – variable in number, and as a rule, numerous (e.g., more than twice as many stamens as petals or sepals), but no standard number of stamens. In another usage, it is a synonym for the preferable term *indeterminate*, meaning the condition in which a flower does not terminate an inflorescence but continues growing until limited by physiological factors. Compare *numerous*. Contrast *definite*.

Indehiscent – not opening in any definite manner at maturity; usually referring to the fruit. Contrast *dehiscent*.

Indeterminate – usually referring to a stem or inflorescence in which no terminal bud or meristem stops growth and ends the extension of the stem, which continues until physiological factors stop the growth. Racemes of some *Xanthorrhoeaceae*, such as many Aloes, and many *Iridaceae*, such as Watsonias, are indeterminate. Contrast *determinate*.

Indigenous – native to the area, not introduced, and not necessarily confined to the region discussed or present throughout it (like '*native*' but usually applied to a smaller area). For example, the Cootamundra Wattle is native to Australia but indigenous to the Cootamundra region of southern New South Wales. Compare *endemic*.

Indumentum – a collective term for a surface covering of trichomes (e.g., hairs, scales).

Induplicate – folded upward or folded with the two adaxial surfaces together.

Indusia– covers groups of sporangia (sori).

Any duplication (copies, uploads, PDFs) is illegal.

Inferior ovary – most of the flower is attached to the top of the ovary. The inferior ovary only corresponds with monomeric or syncarpous flowers. At least partly below the level of attachment of other floral parts. Compare the *superior ovary* and *half-inferior ovary.*

Inflated – swollen, like a bladder.

Inflexed – bent sharply upward or forward. Compare *deflexed.*

Inflorescence – isolated generative shoot. Several flowers are closely grouped to form an efficiently structured unit: the grouping or arrangement of flowers on a plant.

Infraspecific – denotes taxonomic ranks below species level, such as subspecies.

Infrageneric – denoting taxonomic ranks below the genus level, for example, subgenera, sections, and series.

Infructescence – the grouping or arrangement of fruits on a plant.

Infundibular (or *infundibuliform*) – funnel-shaped, for example, in a flower's corolla.

Inrolled – rolled inward.

Insectivorous – catching and drawing nutriment from insects.

Insertion (*point of*) – the point at which one organ or structure (e.g., a leaf) is joined to the structure that bears it (e.g., a stem).

Inserted – growing out from.

Integument – an extra cover of megasporangium. In general, any covering, but especially the covering of an ovule.

Intercalary (*of growth*) – occurring between an organ's apex and base.

Intercalary meristems – grow in two directions and between the apex and the base of an organ.

Interjugary glands – in pinnate leaves, glands occurring along the leaf rachis between the pinnae (occurring below the single, and often slightly larger, gland at or just below the insertion of the pinnae). Compare *jugary.*

Internode – spaces between nodes. *The portion of a stem between two nodes.*

Interpetiolar (*of stipules*) – between the petioles of opposite leaves (e.g., in *Rubiaceae*).

Intramarginal – inside but close to the margin, for example, a vein in a leaf.

Intrastaminal – inside the stamens or androecium, usually referring to the location of a nectary disk.

Introrse – of anther locules, with an opening toward the center of the flower (at least in the bud). Compare *extrorse* and *latrorse.*

 Copyright © Sterling Education

Invalid – using names not validly published according to the Code (i.e., not strictly 'names' in the sense of the International Code of Botanical Nomenclature).

Involucre – a structure surrounding or supporting, usually a head of flowers. In *Asteraceae*, the phyllaries (bracts) surround the inflorescence before opening, supporting the cup-like receptacle on which the head of flowers sits. In *Euphorbiaceae*, the cuplike structure holds the nectar glands, nectar, and head of flowers and sits above the bract-like cyathophyll structure. Involucres occur in Marchantiophyta, Cycads, fungi, and many other groups.

Involute – rolled inward, for example, when the margins of a leaf are rolled toward the adaxial (usually upper) surface. Compare *revolute*.

Iridescent – having a reflective colored sheen produced by structural coloration, as in the speculum of the mirror orchid *Ophrys speculum*.

Irregular – not able to be divided into two halves through any vertical plane. See *asymmetrical*. Compare *zygomorphic*, *actinomorphic*, and *regular*.

Isidium (pl. *isidia*) – a warty club-like structure in some lichens that breaks off and forms new lichens without sexual reproduction. Isidia are dispersed by mechanical means, compared to *soredia*, which are dispersed by wind.

Isobifacial (*of flat structures, especially leaves*) – having both surfaces similar, usually referring to cell types or the number and distribution of stomata.

Isomerous – having an equal number of parts in the whorls.

Isotomic – having branches of equal diameter. Compare *anisotomic*.

J

Joint – a node or junction of two parts; articulation.

Jugary – associated with a jugum or something yoke-like; see, for example, the *jugary gland*.

Jugary gland – occurring on the rachis of a pinnate or bipinnate leaf on a jugum, the junction or attachment of pairs of pinnae or pinnules, as in some Acacia species. Compare *interjugary*.

Jugate – yoke-like; describing a structure of paired items joined together as in a jugum or something yoke-like, such as some leaves and fruit.

Jugum – applied to various yoke-like organs, usually in the sense of their being paired, such as a pair of pinnae on a rachis.

Juvenile leaves – formed on a young plant and typically differ from adult leaves in form.

Any duplication (copies, uploads, PDFs) is illegal.

K

K-strategy – population growth when there is a small number of offspring with a high probability of survival.

Keel – the petal, where the bottom two petals, below the wings, in flowers of the family *Papilionaceae* of the flowering plant. See carina.

Krebs Cycle (or *Citric Acid Cycle*) – occurs in the mitochondrion after glycolysis. It is the second of three phases of cellular respiration; it produces ATP molecules, energy (NADH and $FADH_2$) carrier molecules, and CO_2 from each glucose molecule.

L

Labellum – lip; one of three or five petals is (usually) different from the others.

Labiate – lipped; where a corolla is divided into two parts, called an upper and lower lip, the two resembling an open mouth with lips.

Lacerate – jagged, as if torn.

Lacinia (*foliose lichens*) – a linear to elongate lobe, usually arising from or at the end of a larger lobe

Laciniate (*of lobes*) – with ends irregularly divided into deeply divided, narrow, pointed segments; Of margins – deeply divided into pointed segments irregularly.

Lactic acid fermentation – occurs in animal cells when no oxygen is available.

Lateral meristem cambium – meristem appearing sideways.

Lateral roots – extend horizontally from the tap root and act to anchor the plant.

Lateral veins – smaller veins, typically branching out of the main vein.

Lacuna – a space, hole, cavity, pit, depression, or discontinuity.

Lamella – a thin, plate-like layer. (plural lamellae; adjective lamellate – composed of an assemblage of many layers)

Lamina – the blade of a leaf or the expanded upper part of a petal, sepal or bract.

Laminal – of, or pertaining to, the upper surface, such as the cortex of a lichen.

Lanate – covered in or composed of woolly hairs.

Lanceolate – longer than broad, narrowly ovate, broadest in the lower half, and tapering to the tip, like a lance or spearhead (sometimes, and incorrectly, used to mean narrowly elliptic).

Lateral – attached to the side of an organ (e.g., leaves or branches on a stem). See *dorsal*.

Latex – a milky fluid that exudes from plants such as spurges, figs, and dandelions.

 Copyright © Sterling Education

Laticiferous – latex-bearing, producing a milky juice.

Latrorse – an anther dehiscence in which the anthers open laterally toward adjacent anthers. Compare *introrse* and *extrorse*.

Lauroid – resembling *Laurus*, the laurel genus, particularly its leaves.

Lax – loose, not compact. Of bundles of hyphae in lichens; not stiff and not adglutinate.

Leaf – a flat, usually green part of a plant attached to a stem. The outgrowth of a stem is usually flat and green; its main function is food manufacture by photosynthesis.

Leaf gap – a parenchymatous area in the stele above (i.e., distal to) a leaf trace.

Leaf lateral – photosynthetic organ of a shoot with restricted growth.

Leaf primordia – embryonic leaves.

Leaf scars – 1. marks of leaf petioles. 2. a healing layer forming on a stem where a leaf has fallen off.

Leaf trace – marks of leaf vascular bundles; a vascular bundle connecting the stele to a leaf.

Leaflets – the ultimate segments of a compound leaf.

Leaves – the principal organs of photosynthesis, produce food for the tree.

Lecanorine (*of lichens*) – apothecia with rims of tissue similar to thallus tissue (e.g., genus Lecanora).

Legume – 1. a fruit characteristic of family Fabaceae, formed from one carpel and dehiscent along both sides or indehiscent. 2. a crop species in the family Fabaceae. 3. a plant of the family Fabaceae.

Lemma – the lower of 2 bracts enclosing a grass flower.

Lenticel – typically lenticular (lens-shaped) porous tissue in the bark with large intercellular spaces that allow direct gas exchange between the internal tissues and atmosphere through the bark.

Lenticular – 1. lens-shaped. 2. covered in lenticels.

Lepidote – covered with small scales.

Leprose – powdery.

Leptosporangiatae – sporangia with a one-celled wall.

Liana (or (*liane*)– a woody climbing plant rooted in the ground.

Liane – a woody climbing plant rooted in the ground. See *liana*.

Lichenicolous – growing on or in lichens, often but not necessarily as parasites.

Light-dependent reactions – occur in the thylakoids when energy is captured from sunlight. *Water is split into Hydrogen ions, electrons, and oxygen (O_2).*

Any duplication (copies, uploads, PDFs) is illegal.

Light-independent reactions – occur in the stroma, a series of reactions referred to as the Calvin Cycle. The chemical energy stored in ATP and NADPH powers the formation of Organic Compounds (Sugars) using carbon dioxide (CO_2) and the H^+ from $NADPH^+$.

Ligneous – having hard lignified tissues or woody parts, woody.

Lignum – dead wood, typically in the context of a substrate for lichens.

Lignicolous – grows on wood tissue after the bark has fallen or been stripped off (compare *corticolous*).

Lignotuber – a woody swelling of the stem below or just above the ground; contains adventitious buds from which new shoots can develop (e.g., after fire).

Ligulate – 1. bearing a ligule. 2. strap shaped.

Ligule – 1. a small membranous appendage on the top of the sheath of grass leaves. 2. a minute adaxial appendage near the base of a leaf (e.g., Selaginella). 3. an extended, strap-like corolla in some daisy florets.

Linea (or *line, British line,* and *Paris line*) – pre-metric units larger than 2 mm, used in botany into the 20th century. See *Line* (unit) and *Paris line*.

Linear – very narrow in relation to its length, with the sides mostly parallel. See *leaf shape*.

Lithophytic – a plant growing on rocks; an epilithic plant.

Lobe – part of a leaf (or other organ), often rounded and formed by incisions to about halfway to the midrib.

Loculicidal (*of a fruit*) – dehiscing through the centers of loculi. Compare *septicidal*.

Locule – a chamber or cavity containing seeds within an ovary, pollen within an anther or spores in a sporangium.

Lodicule – one of the two or three-minute organs at the base of the ovary of a grass flower, representing parts of a strongly reduced perianth.

Lomentum (or *loment*) – a pod-like indehiscent fruit that develops constrictions between the segments and, at maturity, breaks into one-seeded segments instead of splitting them open.

Longicidal (*of anthers*) – opening lengthwise by longitudinal slits. Compare *poricidal*.

Lumber – the wood of trees prepared for use as building material.

Lumen – the cavity bounded by a plant cell wall.

Lunate – crescent-shaped.

Luternode – the space between nodes on a stem.

Lyrate – lyre-shaped; deeply lobed, with a large terminal lobe and smaller lateral ones.

 Copyright © Sterling Education

M

Maculate – spotted; marked with spots.

Main vein – the central, most visible vascular bundle of a leaf (*midrib*).

Male flower – see *staminate flower*.

Mallee – a growth habit in which several woody stems arise separately from a lignotuber; a plant with such a growth habit (e.g., many Eucalyptus species); vegetation characterized by such plants.

Mangrove – any shrub or small tree growing in salt or brackish water, usually characterized by pneumatophores; any tropical coastal vegetation characterized by such species.

Margin – the edge of a structure, as in the edge of a leaf blade.

Marginal – occurring at or very close to a margin.

Marsh – a waterlogged area or swamp.

Mast – edible fruit and nuts produced by woody species of plants (e.g., acorns and beechmast) consumed on the ground by wildlife species and some domestic animals.

Maturation zone – root: the oldest part of the root.

Mealy – covered with coarse, floury powder.

Medulla – 1. (*lichen*) the typically undifferentiated tissue underneath the thallus cortex or between the upper and lower cortex if both are present. The medulla is analogous to the tissues underneath a leaf's epidermis (i.e., skin). The uppermost region commonly contains most of the photobionts. 2. pith. See *medullary rays* in wood.

Megaphyllous – with leaves originating from joint branches.

Megasporangia – female sporangia.

Megasporangium – the larger of two kinds of sporangium produced by heterosporous plants, producing large spores containing female gametophytes. Compare *microsporangium*.

Megaspore – female spore; the larger of two kinds of spores produced by a heterosporous plant, giving rise to the female gametophyte. Compare *microspore*.

Megasporophylls (*in heterosporous plants*) – modified leaves with attached megasporangia, a modified leaf bearing one or more megasporangia. Compare *microsporophyll*.

Megastrobilus – the larger of two kinds of cones or strobili produced by gymnosperms, being female and producing the seeds. Compare *microstrobilus*.

Membranous – thin, translucent and flexible, seldom green.

Any duplication (copies, uploads, PDFs) is illegal.

Mericarp – one segment of a fruit (a schizocarp) that splits at maturity into units derived from the individual carpels, or a carpel, usually 1-seeded, released by the break-up at maturity of a fruit formed from 2 or more joined carpels.

Meristem – – specialized groups of cells that are a plant's growing points, cell division sites. *Any actively dividing plant tissue.*

Merosity – multiple flower parts.

Mesic – moist, avoiding both extremes of drought and wet; pertaining to conditions of moderate moisture or water supply; applied to organisms (vegetation) occupying moist habitats.

Mesocarp – the fleshy portion of the wall of a succulent fruit inside the skin and outside the stony layer (if any), surrounding the seed; sarcocarp.

Mesomorphic – soft and with little fibrous tissue, but not succulent.

Mesophyll – 1. the parenchyma tissues between the upper and lower epidermis. They vary in function but usually include the photosynthetic tissue of a leaf. *Photosynthetic parenchyma of the leaf. A leaf's inner tissue between the upper and lower epidermis contains chloroplasts and other specialized cellular parts (organelles).*2. (*in ecology*) the blade of a leaf or leaflet that has a surface area of 4500–18225 mm^2; a plant, or vegetation, that has mesophyll (sized) leaves.

Mesophyllous – (*of vegetation*) Of moist habitats and having mostly large and soft leaves.

Mesophyte – plants adapted to the average water. A plant thriving under intermediate environmental conditions of moderate moisture and temperature without major seasonal fluctuations.

Micropyle – opening at the apex of an ovule.

Microsporangium – the smaller of two kinds of sporangium produced by a heterosporous plant, producing microspores that contain the male gametophyte. Compare *megasporangium.*

Microspore – male spores. The smaller of two kinds of spores produced by a heterosporous plant. Compare *megaspore.*

Microsporgangia –male sporangia.

Microsporophyll (*in heterosporous plants*) – a modified leaf bearing one or more microsporangia. Compare *megasporophyll.*

Microstrobilus – the smaller of two kinds of cones or strobilus produced by gymnosperms, being male and producing the pollen. Compare *megastrobilus.*

Midrib (or *midvein*) – the central and usually most prominent vein of a leaf or leaf-like organ.

Midvein – see *midrib.*

Mitochondria – organelle where cellular respiration occurs.

Copyright © Sterling Education

Monad – a single individual free from other individuals and not united into a group. The term is usually used for pollen to distinguish single grains from tetrads or polyads.

Moniliform – resembling a string of beads.

Monilophytes – all Pteridophyta except lycophytes.

Monocarpic – flowering and setting seed only once before dying. See *semelparous*.

Monochasium – a cymose inflorescence with the branches arising singly. Compare *dichasium* and *pleiochasium*.

Monocot (or *monocotyledon*) – having one seed leaf.

Monocotyledon – a flowering plant whose embryo contains one cotyledon (seed-leaf). Compare *dicotyledon*.

Monoecious (*of vascular plants*) – hermaphroditic, unisexual flowers with both sexes on the same plant., or with male and female reproductive structures in separate flowers but on the same plant, or of an inflorescence that has unisexual flowers of both sexes. Contrast *dioecious*.

Monoicous (*of bryophyte gametophytes*) – hermaphroditic or bisexual, where both male and female reproductive structures develop on the same individual. Contrast *dioicous*.

Monograph – of a group of plants, a comprehensive treatise presenting an analysis and synthesis of taxonomic knowledge of that taxon; the fullest account possible (at the time) of a family, tribe, or genus. It is generally worldwide in scope and evaluates all taxonomic treatments of that taxon, including studies of its evolutionary relationships with other related taxa and cytological, genetic, morphological, palaeobotanical, and ecological studies. The term is often incorrectly applied to any systematic work devoted to a single taxon. Compare *revision*.

Monomorphic – one type rather than several. See *dimorphic* (two) and polymorphic (many types).

Monophyllous – having a single leaf.

Monopodial – branching: when the terminal bud continues to grow every year. A mode of stem growth and branching in which a single dominant meristem forms the main axis. Contrast *sympodial*.

Monostromatic – being a single cell thick, as in the alga *Monostroma*.

Monotypic – containing only one taxon of the next lower rank (e.g., a family with only one genus or a genus that includes only a single species).

Morphology – visible, external structure. The shape or form of an organism or part thereof.

Mucro (or *mucronule*) – a sharp, short point, generally at the tip of a leaf or the tip of the midrib of a compound leaf.

Mucronate – terminating in a *mucro*.

Any duplication (copies, uploads, PDFs) is illegal.

Multiple fruit – originated from many pistils. A cluster of fruits produced from more than one flower and appearing as a single fruit, often on a swollen axis, as with many species of the family *Moraceae*. Compare *aggregate fruit*.

Muricate – covered with short, hard protuberances.

Mutation – historically, a mutation was regarded as an abrupt, sometimes heritable, variation from the norm of a population. For example, a plant might unexpectedly produce "double" flowers, a novel color, or a habit of growth uncharacteristic of the species or variety. Advances in genetics and molecular biology in the mid-twentieth century showed that biological mutations comprise and reflect changes in the nucleic acid molecules that encode the genome of an organism or virus. The nucleic acid affected could be chromosomal or extrachromosomal DNA (typically DNA in the mitochondria or chloroplasts). In RNA viruses, a mutation would be a change to the genetic information that the RNA encodes.

Mycelium – the *vegetative* (i.e., nonreproductive) part of a fungus, mostly composed of aggregations of hyphae. It functions in substrate decomposition and absorption of nutrients.

Mycobiont – the fungal component of a lichen. Compare *photobiont*.

Mycoparasites – plants feeding on soil fungi.

Mycorrhiza (pl. *mycorrhizae;* adj. *mycorrhizal*) – roots symbiotic with fungi. One of several symbiotic associations between a fungus and the roots of a plant.

Mycotroph (adj. *mycotrophic*) – a plant that obtains most or all its carbon, water, and nutrients by associating with a fungus.

N

Native – naturally occurring in an area but not necessarily confined to it. Compare *endemic*.

Natural hybrid – a hybrid taxon produced by chance in the wild.

Naturalized – describing a plant, introduced from another region, that grows and reproduces readily in competition with the natural flora.

Nectar – a usually sweet, nutrient-rich fluid produced by the flowers of many plants and collected by bees and other insects.

Nectary (adj. *nectariferous*) – a specialized gland that secretes nectar.

Neophyte – a plant that has recently been introduced to a geographic area. Contrast *archaeophyte*.

Nerve (or *vein*) – a strand of vascular tissue (e.g., in the leaves of vascular plants).

Node – an area on a stem where a leaf, stem, or flower bud attaches. The part of a stem from which leaves or branches arise.

 Copyright © Sterling Education

Nomenclature – the naming of things; often restricted to the correct use of scientific names in taxonomy; a system that sets out provisions for the formation and use of names.

Noxious – of plants containing harmful or unwholesome qualities. Applied with 'weed' to describe a plant that legislation deems harmful to the environment. *For example, each state has specific legislation governing noxious weeds.*

Nucellus – wall of megasporangium. *The tissue of the ovule of a seed plant that surrounds the female gametophyte. Integuments enclose it and are not of epidermal origin.*

Numerous – stamens are described as numerous when there are more than twice as many sepals or petals, especially when no set number exists. Compare *indefinite*.

Nut – a hard, dry, indehiscent fruit containing only one seed.

Nutlet – 1. a small nut. 2. one of the lobes or sections of the mature ovary of some *Boraginaceae*, *Verbenaceae*, and *Lamiaceae* members.

O

Ob- (*prefix*) – meaning *inversely*; usually the same shape as that described by the word stem but attached by the narrower end. See *obcordate*, *oblanceolate* and *obovate*.

Obconic (*of a fruit, hypanthium, pistil, or calyx*) – shaped like an inverted cone attached at the apex.

Obcordate (*of a leaf blade*) – broad and notched at the tip; heart-shaped but attached at the pointed end.

Obdiplostemonous – having stamens arranged in two whorls and twice as many stamens as petals, with the outer whorl opposite the petals. Compare *diplostemonous*, and *haplostemonous*.

Oblanceolate – has a lanceolate shape but broadest in the upper third.

Obligate (*of parasites*) – unable to survive without a host. Contrast *facultative*.

Oblique – slanting of a leaf or stem, larger on one side of the midrib than the other, asymmetrical.

Obolid – having a three-dimensional oblong shape (e.g., fruit).

Oblong – having a length a few times greater than the width, with sides almost parallel and ends rounded.

Obovate (*of a leaf*) – having a length about 1.5 times the width and widest above the center.

Obsolete – not evident, or at most rudimentary or vestigial.

Obtrapeziform – trapeziform, but attached by the narrower trapezoidal base (e.g., of a leaf)

Any duplication (copies, uploads, PDFs) is illegal.

Obtuse – blunt or rounded; having converging edges that form an angle of more than 90°. Compare *acute*.

Ocrea (or *ochrea*) – part of the leaf which goes upwards along the stem. *The sheath formed from two stipules encircling the node in members of the Polygonaceae.*

Odd-pinnate (or *imparipinnate*) – having an odd number of leaflets in a compound pinnate leaf, with only one terminal leaflet.

Ontogeny – the sequence of developmental stages an organism passes as it grows.

Operculum (*calyptra*) – a lid or cover that becomes detached at maturity (e.g., in *Eucalyptus*); a cap covering the bud and formed by the fusion or cohesion of perianth parts.

Opposite – 1. describing leaves or flowers borne at the same level but directly opposite sides of their common axis. 2. describing the occurrence of something on the same radius as something else (e.g., anthers opposite sepals). Compare *alternate*.

Opposite leaf arrangement – two leaves per node.

Orbicular – flat and more or less circular.

Order – a group of one or more families sharing common features, ancestry, or both.

Organ – the union of different tissues with a common function and origin.

Ortet – the original single-parent plant from which a clone ultimately derives.

Orthotropic – vertical growth.

Orthotropous – when an ovule is erect, with the micropyle directed away from the placenta; atropous. Compare *amphitropous*, *anatropous*, and *campylotropous*.

Oval – see *elliptical*.

Ovary – the basal portion of a carpel or group of fused carpels enclosing ovules. Part of a female flower where eggs are located. *The ovary develops into fruit; ovules develop into seeds after fertilization.*

Ovate – shaped like a section through the long axis of an egg and attached by the wider end.

Ovoid – egg-shaped, with a wider portion at the base; 3-dimensional object, ovate in all sections through the long axis.

Ovule – seed plants: megasporangium with integument. Loosely, the seed, before fertilization, is a structure in a seed plant within which one or more megaspores are formed (after fertilization, it develops into a seed).

Oxygen debt – occurs during anaerobic respiration when oxygen must be 'paid back' to cells to remove lactic acid.

Oxylophytes – plants adapted to acidic substrates.

 Copyright © Sterling Education

P

Pachycaul – with a disproportionately thick trunk

Pachycladous – with disproportionately thick stems

Palea *(plural paleae)* – 1. the upper of two bracts enclosing a grass flower, major contributors to chaff in harvested grain. 2. chaffy scales on the receptacles of many *Asteraceae*. 3. chaffy scales on the stipe of many ferns.

Paleate – bearing paleae or chaffy scales, as in the description of the receptacle of a capitulum (e.g., *Asteraceae*).

Paleaceous – chaff-like in texture.

Palisade mesophyll – mesophyll of elongated, tightly packed cells.

Palmate – 1. leaf with veins radiating from a central point (usually at the top of a petiole), resembling spread-out fingers pointing away from the palm. 2. a compound palmate leaf has leaflets that radiate from a central point (usually at the top of a petiole).

Palmatifid – deeply divided into several lobes arising from more or less the same level.

Palmatisect – intermediate between palmate and palmatifid (i.e., the segments are not fully separated at the base); often more or less digitated.

Panicle (adj. *paniculate*) – a compound raceme; an indeterminate inflorescence in which the flowers are borne on branches of the main axis or further branches of these.

Papilionate – butterfly-like; it has a corolla like that of a pea.

Papilla (pl. *papillae;* adj. *papillose* or *papillate*) – a small, elongated protuberance on the surface of an organ, usually an extension of one epidermal cell.

Pappus (*in daisy florets*) – a tuft or ring of hairs or scales borne above the ovary and outside the corolla (representing the reduced calyx); a tuft of hairs on a fruit.

Paraperigonium (or *paraperigone*) – an anomalous secondary outgrowth of the perianthal meristem with ramifying vasculature. See *perigonium*, *perianth*, and *corona*.

Parasite – an organism living on or in a different organism from which it derives nourishment. Some plant species are parasitic. Compare *saprophyte* and *epiphyte*.

Parcellate – reproduce vegetatively with easily rooted body parts.

Parenchyma – a tissue or cell type of spherical, roughly connected living cell. A versatile ground tissue composed of living primary cells that perform various structural and biochemical functions in plants.

Parietal – attached to the marginal walls of a structure (e.g., ovules attached to placentas on the wall of the ovary). See *placentation*.

Any duplication (copies, uploads, PDFs) is illegal.

Paripinnate – having an even number of leaflets or pinnae (i.e., terminated by a pair of pinnae instead of a single pinna). Compare *imparipinnate.*

Parthenocarpy – the development or production of fruit without fertilization. Compare *stenospermocarpy.*

Patent (or *patulous*) – spreading; standing at 45–50° to the axis. See *erecto-patent.*

Patulous – see *patent.*

Pauciflor – having few flowers per inflorescence. Compare *pluriflor* and *uniflor.*

Pectinate – pinnately divided into narrow segments closely set like the teeth of a comb.

Pedate – having a terminal lobe or leaflet, and on either side of it an axis curved outward and backward, bearing lobes or leaflets on the outer side of the curve.

Pedicel (adj. *pedicellate*) – the stalk of a flower; may be applied to the stalk of a capitulum in the *Asteraceae.*

Peduncle (adj. *pedunculate*) – the stalk of an inflorescence.

Peltate – shield-like, with the stalk attached to the lower surface, not the margin.

Pellucid – transmitting light; for example, said of tiny gland dots in the leaves (e.g., *Myrtaceae* and *Rutaceae*) that are visible when held in front of a light.

Pendulous – hanging, for example, an ovule attached to a placenta on the top of the ovary. Compare *suspended.*

Penicillate – tufted like an artist's brush, with long hairs toward one end.

Penninervation (adj. *penninerved*) – with pinnately arranged veins.

Pentamerous – in five parts, particularly concerning flowers, five parts in each whorl. See *trimerous* and *tetramerous.*

Pepo – berry formed from an inferior ovary containing many seeds, usually large with a tough outer skin (e.g., a pumpkin or cucumber).

Perennating – an organ that survives vegetatively from season to season. *A period of reduced activity between seasons is usual.*

Perennial – a plant whose life span extends over several years.

Perfect (*of a flower*) – bisexual; containing both male and female reproductive parts in the same inflorescence. Contrast *imperfect.*

Perfoliate – with its base wrapped around the stem so that the stem appears to pass through (e.g., leaves and bracts).

 Copyright © Sterling Education

Perforate – with many holes. Used to describe the texture of pollen exine and to indicate that tracheary elements have a perforation plate. See *fenestrate*.

Perforation plate – in a tracheary element, part of the perforated cell wall, present in vessel members but not in tracheids. It should not be confused with a *pit*.

Perforations – openings.

Perianth – the collective term for the calyx and corolla of a flower (generally used when the two are too similar to be easily distinguishable). Abbreviation: *P*; for instance, P 3+3 indicates the calyx and corolla each have 3 elements (i.e., 3 sepals + 3 petals).

Pericarp – most of the fruit tissue. The wall of a fruit develops from the ovary wall.

Periclinal – curved parallel to a surface. Compare *anticlinal*.

Pericycle – parenchyma layer just outside of vascular tissues. A cylinder of parenchyma or sclerenchyma cells lies just inside the endodermis and is the outermost part of the stele of plants.

Periderm – secondary dermal tissue.

Perigonium – 1. in flowering plants, synonym of *perianth*. 2. in mosses, the leaves surrounding the antheridia are also called a *splash-cup* (e.g., in *Polytrichum juniperinum*).

Perigynium – a sac from a modified tubular bract or a utricle around the pistillate flower of sedges when fully closed.

Perigynous – borne around the ovary; perianth segments and stamens arising from a cup-like or tubular extension of a receptacle (free from the ovary but extending above its base). Compare *epigynous* and *hypogynous*.

Perisperm – nutrition tissue originated from nucellus.

Persistent – remaining attached to the plant beyond the usual time of falling, for instance, sepals not falling after flowering, flower parts remaining through fruit maturity. Compare *deciduous* and *caducous*.

Peristome – mosses: attachment to moss sporangium, helps to distribute spores.

Perule (adj. *perulate*) – 1. the scales covering a leaf or flower bud or a reduced scale-like leaf surrounding the bud. Buds lacking perulae are referred to as "naked." 2. (*in Camellias*) the final bracts and sepals become indistinguishable and are called perules. 3. a sac formed by the adherent bases of the two lateral sepals in specific orchids.

Petal – in flower, one of the segments or divisions of the inner whorl of non-fertile parts surrounding the fertile organs, usually soft and conspicuously colored. Compare *sepal*.

Petaloid – like a petal; soft and colored conspicuously.

Petiolary (*or petiolar*) – associated with a petiole, as in petiolary glands.

Petiolate (*of a leaf*) – having a petiole. Contrast *sessile*.

Any duplication (copies, uploads, PDFs) is illegal.

Petiole – the stalk that attaches a leaf to a stem.

Petiolule – the stalk of a leaflet.

Petricolous – rock-dwelling; living on or among rocks.

Phanerogam – gymnosperms and angiosperms; plants producing stamens and gynoecia; plants with conspicuous sexual reproductive organs. Compare *cryptogams*.

Phellem – the external layer of the periderm, cork.

Phelloderm – the internal layer of the periderm.

Phellogen – cork cambium, lateral meristem making periderm.

Phenology – the study of the timing of seasonal biological phenomena, such as flowering, leaf emergence, fruit ripening, and leaf fall.

Phloem – photosynthate-conducting tissue that carries synthesized food substances (i.e., sugars) from the leaves to the rest of the tree and roots. *Vascular tissue transporting sugars. A specialized conducting tissue in vascular plants that transports sucrose from the leaves to other plant organs.*

Photobiont (or *phycobiont*) – in a lichen, the component that does the photosynthesis is green algae (Chlorophyta) or blue-green algae (Cyanobacteria). Compare *mycobiont* (i.e., the fungal component).

Photolysis – the splitting of a water molecule; light energy is converted to chemical energy, which is temporarily stored in ATP and NADPH. Electrons that absorb the energy are passed along the (electron transport chain) ETC and stored in ATP. NADP picks up the H^+ to form $NADPH^+$ and electrons from photolysis and stores it for later use.

Photosynthate – food product (sugar or starch) created through photosynthesis.

Photosynthesis – the process by which green plants use sunlight energy to convert atmospheric CO_2 and H2O from the soil into sugar (e.g., glucose) and oxygen. A series of complex chemical reactions in plants, some algae, and bacteria whereby carbon dioxide and water are combined to form glucose and oxygen; the process requires light and chlorophyll. Energy from sunlight is used to convert CO_2 and water into sugars in cells containing chloroplasts; all plants, except certain parasites, can perform photosynthesis. Photosynthesis equation: $6\ CO_2 + 6\ H_2O + (\text{LIGHT ENERGY}) \rightarrow C_6H_{12}O_6 + 6\ O_2$

Phycobiont – in a lichen, a synonym for photobiont.

Phyllary – an individual bract within an involucre or involucel.

Phyllid – a leaf-like extension of the stem in Bryophytes

Phyllode (adj. *phyllodineous*) – leaf-like petioles. A leaf with the blade much reduced or absent and in which the petiole and or rachis perform the functions of the whole leaf (e.g., many acacias). Compare *cladode*.

Phyllopodium (*in ferns*) – a short outgrowth of the stem on which the frond is borne and remains attached to the rhizome after the frond has been shed.

 Copyright © Sterling Education

Phylloplane – the surface of a leaf, considered as a habitat for organisms.

Phyllosphere – the above-ground surface of plants as a habitat for epiphytic microorganisms.

Phyllotaxis – leaf arrangement.

Phylum –A classification or taxonomic rank below kingdom and above class. Traditionally, the term division has been used instead of phylum in botany.

Phytomelan (or *phytomelanin;* adj. *phytomelanous*) – a black, inert, organic material that forms a crust-like covering of some seeds, commonly found in *Asparagales*, *Asteraceae*, etc.

Pigment – a molecule that absorbs specific wavelengths of light and reflects or transmits others.

Pileate – having a cap, a pileus.

Pileus – a cap or cap-shaped structure, such as the cap of mushrooms or the plumule of some monocotyledons.

Piliform – having the shape of a cap, a pileus.

Pilose – covered with soft, weak, thin and separated hairs, which are usually defined as long and sometimes ascending.

Pinna (pl *pinnae*) – a primary segment of a compound leaf.

Pinnate – a compound leaf with leaflets arranged on each side of a common petiole or axis; also applied to how the lateral veins are arranged about the main vein.

Pinnatifid – pinnately lobed.

Pinnatisect – pinnately divided almost to midrib but segments still confluent.

Pinnule (or *pinnula*) – 1. ultimate free division (or leaflets) of a compound leaf. 2. a pinnate subdivision of a multipinnate leaf.

Pistil – collective term for carpel. *Carpel and pistil are equivalent when there is no fusion; if fusion occurs, two or more carpels unite into one pistil.* 1. a single carpel, when the carpels are free, additional carpels cover ovules. The female flower part consists of a stigma, style, and ovary. 2. a group of carpels when the fusion of their walls unites the carpels.

Pistillate flower (or *female flower*) – containing one or more pistils but no stamens.

Pistillode – a sterile or rudimentary pistil.

Pit (*in tracheary elements*) – structure connecting tracheids. A section of the cell wall where the secondary wall is missing and the primary wall present. *Pits generally occur in pairs and link two cells.*

Pith – the central region of a stem inside the vascular cylinder; the spongy parenchymatous central tissue in some stems and roots.

Placenta – the tissue within an ovary to which the ovules are attached.

Any duplication (copies, uploads, PDFs) is illegal.

Placentation – the arrangement of ovules inside an ovary. For example, axile, free-central, parietal, marginal, basal, or apical.

Placodioid – the form of a lichen thallus which radiates outward with the ends of the radiating arms peeling up from the substrate but which lacks a cortex on the underside (unlike foliose lichens).

Plagiotropic – horizontal growth.

Plant Breeders Rights (PBR) – these rights, governed by Plant Breeder's Rights Acts, give the plant breeder legal protection over the propagation of a cultivar and the exclusive rights to produce and to sell it, including the right to license others to produce and sell plants and reproductive material of a registered, deliberately bred variety. Cf. UPOV.

Plant Variety Rights (PVR) – governed by the Plant Variety Rights, the registration of new varieties is now governed by Plant Breeders Rights.

Plastochron – the time between successive leaf initiation events.

Pleiochasium (pl. *pleiochasia*) – an inflorescence in which several buds come out simultaneously. Compare *monochasium* and *dichasium*.

Plicate – pleated; folded back and forth longitudinally like a fan, such as the leaves of fan palm species. The concept often appears in specific names in forms such as *Kumara plicatilis* and *Acacia plicata*. Such names are not correctly appropriate but are applied to distichous structures rather than plicate.

–**plinerved** (*of leaves*) – suffix indicates that the main nerves are lateral and arise from a point distinctly above the base of the leaf. Combined with a numerical prefix to form words like 3-plinerved, 5-plinerved, and so on. Such leaves are especially characteristic of the family *Melastomataceae* (e.g., *Dissotis*).

Plumose – like a feather, with fine hairs branching from a main axis.

Plumule – the part of an embryo that gives rise to the shoot system of a plant. *Compare radicle*.

Pluriflor – having many flowers per inflorescence.

Pluriovulate – having many ovules, such as in placentae, carpels, or ovaries.

Pneumatophore – air-catching heliotropic roots. A vertical appendage, aerial at low tide, on the roots of some plants. Pneumatophore functions are unclear but related to gas exchange or root anchoring. Pneumatophores typically occur on mangrove roots, but some versions occur on species of conifers.

Pod – 1. a legume, the fruit of a leguminous plant, a dry fruit of a single carpel, splitting along two sutures. 2. a siliqua and silicula, the fruit of *Brassicaceae*, a dry fruit composed of two carpels separated by a partition.

Poikilohydric – plants that do not save water.

Pollen – powdery mass shed from *anthers* (of angiosperms) or *microsporangia* (of gymnosperms); the microspores of seed plants; pollen grains.

 Copyright © Sterling Education

Pollen-mass – pollen-grains covered by a waxy texture or fine threads into a single body; pollinium (e.g., orchids).

Pollen sac – microsporangium seed plants.

Pollen transmitting tissue – the tissue in the style of a flower through which the pollen tubes grow.

Pollen tube – fungus-like cell which brings spermatia to an egg.

Pollination – transfer of male *gametophytes* (or pollen grains) from *microsporangia* (or pollen sacs) to *megasporangia* (or ovules) or *cupules* (or pistils). The transfer of pollen from a male organ (e.g., an *anther*) to the receptive region of a female organ (e.g., a *stigma*).

Pollinium – see *pollen mass*.

Polygamodioecious – having bisexual and male flowers on some plants and bisexual and female flowers on others. Compare *androdioecious*, *andromonoecious*, *dioecious*, *monoecious*, *polygamomonoecious*, and *polygamous*.

Polygamomonoecious – having male, female, and bisexual flowers on the same plant. Compare *androdioecious*, *andromonoecious*, *polygamodioecious*, and *polygamous*.

Polygamous – having bisexual and unisexual flowers on the same plant.

Polymorphic – of several kinds (regarding shape or size), hence polymorphism. See *monomorphic* (a single type) and *dimorphic* (two types)

Polyploid – with more than two of the basic sets of chromosomes in the nucleus; any sporophyte with cells containing three or more complete sets of chromosomes. Words or numbers with *-ploid* indicate the number of haploid sets of chromosomes (e.g., triploid = 3 sets, tetraploid = 4 sets, pentaploid = 5 sets, hexaploid = 6 sets, and so on).

Polystemonous – having numerous stamens; the number of stamens being at least twice the number of sepals or petals, but not strictly three or four times that number.

Pome – a fruit developed partly from the ovary wall but mostly from hypanthium (e.g., an apple).

Population – 1. all of one or more species within a prescribed area. 2. a group of organisms of one species occupying a defined area and usually isolated to some degree from other similar groups. 3. in statistics, the whole group of items under investigation.

Poricidal – opening by pores, as with the capsule of a poppy or the anthers in several families of plants. Compare *longicidal*.

Posterior – positioned behind or toward the rear. Contrast *anterior*.

Prickle (adj. *prickly*) – a hard, pointed outgrowth from the surface of a plant (involving several layers of cells but not containing a vein); a sharp outgrowth from the bark, detachable without tearing wood. Compare *thorn*.

Any duplication (copies, uploads, PDFs) is illegal.

Primary meristems – make vascular tissues. Intermediate tissues start apical meristems and make primary tissues.

Primary root – originates from embryo root.

Primary species – in lichens, a species reproduces mainly by sexual reproduction rather than vegetative reproduction.

Primary stem – with primary tissues only.

Primary tissues – originate from RAM or SAM (optionally through intermediate meristems).

Primary vein – a single or array of veins conspicuously larger than any others in a leaf. In pinnate venation, the single primary vein is generally in the middle of the leaf; in palmate venation, several such veins radiate from a point at or near the base of the leaf.

Procambium intermediate meristem – developing into cortex, pith and procambium.

Procumbent – spreading along the ground but not rooting at the nodes; not as close to the ground as prostrate.

Pro parte – in part. In nomenclature, it is used to denote that the preceding taxon includes more than one currently recognized entity and that only one of those entities is being considered.

Propagule –1. (*in lichens*) a part of the thallus that has both fungal and algal parts and can break off for vegetative reproduction (e.g., *isidium*, *phyllidium*, *phyllocladium*, or *soredium*). 2. any structure capable of generating a new plant, including seeds, spores, bulbils, etc.

Prophyll – a leaf formed at the base of a shoot, usually smaller than those formed later.

Prostrate – lies flat on the ground, commonly rooting at nodes touching the soil surface.

Protandrous – having male sex organs that mature before the female ones (e.g., a flower shedding pollen before the stigma is receptive). Compare *protogynous*.

Proteranthous – with new leaves appearing before flowers. See *hysteranthous* and *synanthous*.

Prothallus – a gametophyte plant, usually flattened and delicate (e.g., ferns and fern allies).

Protoderm – primary meristem produces epidermis or rhizodermis.

Protogynous – having female sex organs that mature before the male ones (e.g., a flower shedding pollen after the stigma has ceased to be receptive). Compare *protandrous*.

Protonema – mosses: embryonic thread of cells.

Protostele – central xylem surrounded with phloem.

Proximal – near the point of origin or attachment. Compare *distal*.

Pruinose – covered with a powdery, waxy material; having a bloom.

Psammophytes – plants adapted to grow on sandy substrates.

Copyright © Sterling Education

Pseudanthium – a type of inflorescence occurring in the *Asteraceae* and *Euphorbiaceae*, where multiple flowers are grouped to form a flower-like structure, commonly called a head or capitulum.

Pseudo- (*prefix*) – meaning *false, not genuine* (e.g., a pseudo-bulb is a thickened, bulb-like internode in orchids, but not an actual bulb).

Pseudobasifixed (*of an anther*) – connected to the filament of the stamen by connective tissue, which extends in a tube around the filament tip. See *basifixed* and *dorsifixed.*

Pseudostipule – an enlarged, persistent axillary bud scale resembling a stipule; common in *Bignoniaceae*.

Pseudoverticillate – having the appearance of being whorled (i.e., *verticillate*) without being so.

Pteridophytes – plants adapted to grow on rocky substrates.

Puberulous (or *puberulent*). – covered with minute, soft, erect hairs.

Pubescent – downy; covered with short, soft hairs, especially erect hairs.

Pulverulent – having powdery or crumbly particles as if pulverized.

Pulvinate – having a pulvinus.

Pulvinus – a swelling at either end of a leaf's petiole or petiolule of a leaflet (e.g., in *Fabaceae*, that permits leaf movement).

Punctate [Latin, means *puncture* or *prick-mark*) – marked with indefinite dots or similarly small items such as translucent glands or tiny hollows.

Punctiform – dot-like or in the shape of a prick mark.

Pungent – having a sharp, hard point.

Pustule – a blister-like swelling.

Pustulate – having pustules.

Pyramidal (*of a growth habit*) – conical or pyramid-shaped. Most familiar in some coniferous trees, especially species adapted to snowy climates.

Pyrene – the stone of a drupe, consisting of seed surrounded by the hardened endocarp.

Pyriform – pear-shaped; a term for solid shapes that are roughly conical in shape, broadest at one end and narrowest at the other. As a rule, the distal third of their length is the broadest, and they are narrowest near the proximal end, the base, where the stalk, if any, attaches.

Pyrophile – plants that need fire for their reproduction.

Pyrophytes – plants that have adapted to tolerate fire.

Pyruvic acid – converted into a waste product of lactic acid.

Any duplication (copies, uploads, PDFs) is illegal.

Q

Quiescent – center core part of root apical meristem.

R

R-strategy – population growth when there is a vast number of offspring with a low probability of survival.

Raceme – basic monopodial branched inflorescence.

Rachilla (or *rhachilla*) – the axis of a grass spikelet above the glumes.

Rachis (pl. *rachises or rachides*) – the axis of an inflorescence or a pinnate leaf (e.g., ferns; secondary rachis is the axis of a pinna in a bipinnate leaf distal to and including the lowermost pedicel attachment.

Radial – 1. cross-section. 2. with structures radiating from a central point as spokes on a wheel (e.g., lateral spines of a cactus).

Radiate (*of daisies, of a capitulum*) – with ray floret surrounding disc florets.

Radical – springing from the root; clustered at the base of the stem.

Radicle – the part of an embryo giving rise to a plant's root system. Compare *plumule*.

Rainforest – a moist temperate or tropical forest dominated by broad-leaved trees that form a continuous canopy.

RAM (root apical meristem) – an undifferentiated meristematic tissue in the growing tips of plant roots. Its primary function is to trigger the growth of new cells in young seedlings at the tips of roots.

Ramet – an individual member of a clone.

Ramicaul – a single-leafed stem, as in Pleurothallis orchids.

Ray – initials cambium cells that make rays. 1. zygomorphic (ligulate) flowers in a radiate flowerhead (i.e., ray-florets/flowers, e.g., *Asteraceae*). 2. each of the branches of an umbel.

Rays – stem: parenchyma cells arranged for horizontal transport.

Receptacle (or *torus*) – the axis of a flower, in other words, floral axis; torus; e.g., in *Asteraceae*, the floral base or receptacle is the expanded tip of the peduncle on which the flowers are inserted.

Recumbent – bent back toward or below the horizontal.

Recurved – bent or curved backward or downward.

Reduplicate – folded outward or with the two abaxial surfaces together.

Copyright © Sterling Education

Reflexed – bent sharply back or down.

Registered name – a cultivar name accepted by the International Cultivar Registration Authority.

Registration – 1. recording a new cultivar name with an International Cultivar Registration Authority. 2. recording a new cultivar name with a statutory authority like the Plant Breeder's Rights Office. 3. recording a trademark with a trademark office.

Regular – see *actinomorphic*.

Reniform – kidney-shaped.

Repetitive – characters in leaf description that apply to the leaf parts on each hierarchy level.

Replum – a framework-like placenta to which the seeds attach and remain after each valve drops away.

Respiration – the process of converting sugars and starches into energy.

Resupinate – 1. (*in botany*) describes leaves or flowers in an inverted position because the petiole or pedicel, respectively, is twisted 180 degrees. compare: hyper-resupinate. 2. (*in lichenology*) refers to having or being a fruiting body that lies flat on the substrate, with the hymenium either over the whole surface or at the periphery.

Reticulate – forming a network (or *reticulum*). *For example, veins that join one another at more than one point.*

Retrorse – bent backward or downward. Compare *antrorse*.

Retuse – having a blunt (or obtuse) and slightly notched apex.

Revision – an account of a particular plant group, like an abbreviated or simplified monograph. Sometimes, it is confined to the plants of a particular region. Like a monograph in clearly distinguishing the taxa and providing a means for their identification. Compare *monograph*.

Revolute – rolled under (i.e., downward or backward), for example, when the edges of leaves are rolled under toward the midrib. Compare *involute*.

Rheophytes – water plants adapted to fast-moving water.

Rhizine – the "root" or "trunk" projection of a foliose lichen that attaches the lichen to the substrate (what the lichen is growing on)

Rhizodermis – the root epidermis, the outermost primary cell layer of the root

Rhizoid cells – dead cells accumulating water apoplectically.

Rhizome (or *rhiz.*) – 1. underground horizontal shoot. 2. a perennial underground stem usually growing horizontally. See *stolon*.

Rhizomatous – adj. a plant whose above-ground stem is derived from a below-ground stem (rhizome). Compare *arhizomatous* or *arhizomatic*).

Any duplication (copies, uploads, PDFs) is illegal.

Rhizosphere – the below-ground surface of plants and adjacent soil as a habitat for microorganisms.

Rhytidome – the dead region of the bark that lies outside the periderm.

Rhombic – like a rhombus: an oblique figure with four equal sides. Compare *trapeziform*, and *trullate.*

Rhomboid – a four-sided figure with opposite sides parallel but with adjacent sides an unequal length (like an oblique rectangle); see *rhombic.*

Rhomboidal – a shape, for instance, of a leaf, which is diamond-shaped with length equal to width.

Rimose – with many cracks, as in the surface of a crustose, areolate lichen.

Ring – porous wood. *With large vessel elements mostly in early wood.*

Root – an axial organ of a plant with geotropic growth. A unit of a plant's axial system, which is usually underground, does not bear leaves, tends to grow downward, and is typically derived from the radicle of the embryo. *The underground part of a tree attaches it to the ground and draws in water and nutrients.*

Root apical meristem (RAM) – an undifferentiated meristematic tissue in the growing tips of plant roots. Its primary function is to trigger the growth of new cells in young seedlings at the tips of roots.

Root cap – protects root meristem.

Root hairs – outgrowths of the outermost layer of cells just behind the root tips, functioning as water-absorbing organs.

Root microbiome – the dynamic community of microorganisms associated with plant roots.

Root nodules – bulb-like structures that contain nitrogen-fixing bacteria.

Root pressure – a force created solely by roots.

Rootstock (or *stock*) – 1. the part of a budded or grafted plant that supplies the root system. 2. plants selected to produce a root system with some specific attribute (e.g., a virus-free rootstock).

Rosette – when parts are not whorled or opposite but appear so due to contractions of internodes. For example, the petals in a double rose or a basal cluster of leaves (usually close to the ground).

Rostellate – possessing a beak (*rostellum*). Synonym of *rostrate.*

Rostrate – with a beak.

Rotate – circular and flattened; for example, a corolla with a very short tube and spreading lobes (e.g., some *solanaceae*).

Ruderal – a plant that colonizes or occupies disturbed waste ground. See *weed.*

Rudiment – in the structure of a plant, an item that is at best hardly functional, either because it is immature and has not yet completed its development (e.g., a leaf still incompletely formed inside a bud) or because its role in the organism's morphology cannot be completed and therefore is futile (e.g., leaf rudiment at the tip of a phyllode, that will be shed while immature, because the phyllode will take over the leaf function). Compare *cataphyll* and *vestige.*

Copyright © Sterling Education

Rudimentary – being of the nature of a rudiment; at most barely functional because incompletely developed; begun, but far from completed, either temporarily or permanently. Compare *vestigial.*

Rugose – wrinkled, either covered with wrinkles or crumpled like a wrinkled leaf, either as a stiffening structure or in response to disease or insect damage.

Rugulose – finely wrinkled.

Ruminate (*usually applied to endosperm*) – irregularly grooved or ridged and appears chewed (e.g., endosperm in particular members of *Myristicaceae*).

Runcinate – sharply pinnatifid or cleft, with the segments directed downward.

Runner – see *stolon.*

Rush – a plant of the family *Juncaceae* or, more loosely, applied to various monocotyledons.

S

Saccate – pouched or shaped like a sack.

Sagittate – shaped like an arrowhead; narrow and pointed but gradually enlarged at the base into two straight lobes directed downward; may refer only to the base of a leaf with such lobes. Compare *hastate*.

Salverform – trumpet-shaped, having a long, slender tube and a flat, abruptly expanded limb.

SAM (stem apical meristem) – an undifferentiated meristematic tissue in the buds and growing tips of plants. Its primary function is to trigger the growth of new cells in young seedlings at the tips of shoots and forming buds.

Samara – a dry, indehiscent fruit with its wall expanded into a wing (e.g., in the genus *Acer*).

Samphire – a common name given to various edible coastal plants.

Sap – a watery fluid that circulates in a plant. *Sap* is the liquid that flows from the roots to provide water and nutrients to a tree. The *sap* from maple trees is probably the most famous because it becomes maple syrup when boiled down.

Sapwood (or *xylem*) – a specialized water-conducting tissue in vascular plants. Vascular tissue transporting water and nutrient-conducting tissue that carries water and nutrients from the roots to the leaves. Plant tissue that conducts water and dissolved nutrients. *As new layers develop, the inner layers die and become heartwood.*

Saprophyte (adj. *saprophytic*) – a plant, or loosely speaking, a fungus or similar organism, deriving its nourishment from decaying organic matter such as dead wood or humus, and usually lacking chlorophyll. Compare *parasite*, *saprotroph*, and *epiphyte*.

Any duplication (copies, uploads, PDFs) is illegal.

Saprotroph (adj. *saprotrophic*) – an organism deriving its nourishment from decaying organic matter. Contrast *parasite* and *epiphyte*.

Sapwood (or *xylem*) – 1. the pipeline transporting water and nutrients from the roots to the leaves. 2. functional part of wood. *As new layers develop, the inner layers die and become heartwood.*

Sarment – a long, slender, prostrate stolon, commonly called a runner.

Sarmentose – reproducing by sarments; strawberry plants are the most familiar example.

Saxicolous – growing on stone, like some lichens.

Scabrid (or *scabrous*) – rough to the touch, with short, hard protrusions or hairs.

Scalariform – ladder-like in structure or appearance.

Scale – 1. a reduced or rudimentary leaf, for example, around a dormant bud. 2. a flattened epidermal outgrowth, such as those commonly found on the leaves and rhizomes of ferns.

Scandent – climbing, by whatever means.

Scape (adj. *scapose*) – a stem-like flowering stalk of a plant with radical leaves.

Scapose – having the floral axis more or less erect with few or no leaves; consisting of a scape.

Scarious – dry and membranous.

Schizocarp – fruits that segregate into smaller indehiscent units. A dry fruit formed from more than one carpel but breaks apart into individual carpels (mericarps) when ripe.

Scion – the aerial part of a graft combination, induced by various means to unite with a compatible understock or rootstock.

Sciophytes – plants adapted to shade.

Sclereid – a cell with a thick, lignified cell wall that is shorter than a fiber cell and dies soon after the thickening of its cell wall.

Sclerenchyma – dead supportive tissue. A strengthening or supporting tissue composed of sclereids or a mixture of sclereids and fibers.

Sclerophyll (or *sclerophyllous*) – a plant with hard, stiff leaves; any structure stiffened with thick-walled cells.

Sclerophytes – plants preventing water loss often use *sclerenchyma* (i.e., dead supportive tissue).

Scorpioid (*of a cymose inflorescence*) – branching alternately on one side and then the other. Compare *helicoid.*

Scrobiculate – having very small pits.

Scrubland – dense vegetation dominated by shrubs.

 Copyright © Sterling Education

Scurf – minute, loose, membranous scales on the surface of some plant parts, such as leaves.

Secondary metabolites – chemicals produced by plants that do not have a role in so-called primary functions such as growth, development, photosynthesis, reproduction, etc.

Secondary (lateral) roots – originate from primary root (see) secondary vascular tissues.

Secondary species – in lichens, a "*species*" taxon reproduces only by vegetative means, whose components reproduce mainly by sexual means. Contrast *primary species*.

Secretory tissue – concerned with the secretion of gums, resins, oils and other substances in plants.

Section – supplementary taxa intermediate in rank between subgenus and series. It is a singular noun that is always written with a capital initial letter and a generic name.

Secund – having all the parts grouped on one side or turned to one side (e.g., especially for inflorescences).

Sedge – a plant of the family *Cyperaceae*.

Seed – chimeric structure with mother (seed coat), daughter (embryo), and endosperm genotypes. A ripened ovule, consisting of a protective coat enclosing an embryo and food reserves; a propagating organ formed in sexual reproductive cycle of gymnosperms and angiosperms (together, the seed plants).

Seed leaf (or *cotyledonf*) – embryonic leaf. The first leaf that appears on a seedling. The primary leaf or leaves of a plant embryo develop into the seed leaf or the first set of leaves upon germination.

Seed scales – megasporophylls of conifers.

Segment – a part or subdivision of an organ (e.g., a petal is a segment of the corolla); sometimes used when the sepals and petals are indistinguishable.

Self-pollination (or *selfing*) – the acceptance by stigmas of pollen from the same flower or flowers on the same plant, which means they are self-compatible.

Semaphyll – a structure such as a bract or a sepal (if the remainder of the perianth is inconspicuous) modified to attract pollinators.

Semelparity – when a plant flowers once and then dies.

Semiterete – rounded on one side but flat on the other. See *terete*.

Senecioid (or *anthemoid*) (*in Compositae*) – a style with a brush-like tuft of sweeping hairs at the tip of each style branch.

Sensitive – a descriptive term for stigmas that, in response to touch, close the two lobes of the stigma together, ending the receptivity of the stigma, at least for the time that the lobes are closed together (e.g., *Mimulus*).

Sensu – in the sense of.

Any duplication (copies, uploads, PDFs) is illegal.

Sepal (*in a flower*) – one of the segments or divisions of the outer whorl of non-fertile parts surrounding the fertile organs, usually green. Compare *petal.*

Septicidal (*of a fruit*) – dehiscing along the partitions between loculi. Compare *loculicidal.*

Septum (pl. *septa*) – a partition (e.g., the membranous wall separating the two valves of the pod of *Brassicaceae*).

Seriate – arranged in rows.

Sericeous – silky with dense, appressed hairs.

Series – the category of supplementary taxa intermediate in rank between section and species. *It is often used as a plural adjective.*

Serrate – toothed with asymmetrical teeth pointing forward, like the cutting edge of a saw.

Serrulate – finely serrate.

Sessile – attached without a stalk (e.g., a leaf without a petiole or a stigma when the style is absent).

Seta (pl. *setae;* adj. *setose, setaceous*) – mosses: the stalk of sporogony. A bristle or stiff hair (in Bryophytes, the stalk of the sporophyte). A terminal seta is an appendage to the tip of an organ (e.g., the primary rachis of a bipinnate leaf in *Acacia*).

Sheath – part of the leaf which surrounds the stem. A tubular or rolled part of an organ (e.g., the lower part of the leaf in most grasses).

Shoot – the aerial part of a plant; a stem and all its dependent parts (e.g., leaves, flowers).

Shoot plant body – unipolar body: no root system, shoots only.

Shrub – a woody perennial plant without a single main trunk, branching freely, and generally smaller than a tree.

Sieve tube cells – living cells that transport sugar.

Sigmoid – shaped like the letter 'S.'

Silicula (or *silicle*) – a fruit like siliqua, but stouter, not more than twice as long as wide.

Siliqua – a dry, dehiscent fruit (contrast *silicula*, more than twice as long as wide) formed from a superior ovary of two carpels, with two parietal placentas and divided into two loculi by a '*false*' septum.

Silky – densely covered with fine, soft, straight, appressed hairs, with a lustrous sheen and satiny to the touch.

Silviculture – the science of forestry and the cultivation of woodlands for commercial purposes and wildlife conservation.

 Copyright © Sterling Education

Simple – undivided or unsegmented (e.g., a leaf not divided into leaflets). Note that a simple leaf may still be entire, toothed, or lobed) or an unbranched hair or inflorescence.

Simple fruit – originated mainly from one pistil.

Simple leaf – with one level of hierarchy.

Simple tissues – with uniform cells.

Sinuate – having deep, wave-like depressions along the margins, but flat. Compare *undulate.*

Sinus – a notch or depression between two lobes or teeth in the margin of an organ.

Siphonogamy – fertilization with the help of a pollen tube. Fertilization with pollen tubes formed to transfer male gametes to female gametes/eggs. *Most seed plants are siphonogamous. Thus, they accomplish their fertilization by the formation of pollen tubes. Male gametes or sperms of most siphonogamous plants are non-motile.*

Soil – the upper layer of earth that may be a nutrient-rich substrate in which plants grow.

Solenostele – vascular bundles in hollow cylinders.

Solitary – single flowers that grow one plant per year, one in each axil, or widely separated on the plant; not grouped in an inflorescence.

Soralia – in a lichen, the structure that bears *soredium* for non-sexual reproduction.

Soredium (pl. *soredia*) – in a lichen, a small group of algal cells surrounded by fungal filaments form in *soralia,* which break off and grow new lichens without sexual reproduction after being dispersed by wind. Compare *isidium,* which breaks off and is dispersed by mechanical means.

Sorus (pl. *sori*) – a cluster of sporangia. Sori typically occurs in ferns, some algae, and fungi. In many fern species, the sorus is covered by a protective indusium.

sp. – an abbreviation of species (singular), often used when the genus is known but the species has not been determined, as in "*Brassica* sp." See *spp.*

spp. – an abbreviation of species (plural), often used to collectively refer to more than one species of the same genus, as in "*Astragalus* spp." See *sp.*

Spadix – a spicate (spike-like) inflorescence with the flowers crowded densely, even solidly, around a stout, often succulent axis. Particularly typical of the family *Araceae.*

Spathe (adj. *spathaceous*) – a large bract ensheathing an inflorescence. Traditionally, any broad, flat blade.

Spathulate (*or spatulate*) – spoon-shaped; broad at the tip with a narrowed projection extending to the base.

Any duplication (copies, uploads, PDFs) is illegal.

Species – a group, or populations of individuals, sharing common features or ancestry, generally the smallest group that can be readily and consistently recognized; often, individuals capable of interbreeding and producing fertile offspring. *The basic classification unit is the taxa category of the lowest principal rank in the nomenclatural hierarchy. Strict assignment to a species is not always possible, as it is subject to contexts and the species concept under consideration.*

Specific epithet – follows the name of the genus and is the second word of a botanical binomial. The generic name and specific epithet together constitute the name of a species (i.e., the specific epithet is not the species name).

Spermatium – a flagellate, non-motile sperm cell (plural: spermatia).

Spica (adj. *spicate*) – another name for a *spike*.

Spike (adj. *spicate*) – an unbranched, indeterminate inflorescence in which the flowers are without stalks. Compare *raceme*.

Spikelet – a subunit of a spike inflorescence, especially in grasses, sedges, and some other monocotyledons, consisting of one-to-many flowers and associated *bracts* or *glumes*.

Spine (adj. *spinose*) – reduced, prickly leaves. A stiff, sharp structure formed by modifying a plant organ that contains vascular tissue (e.g., a lateral branch or a stipule; includes thorns).

Spinescent – ending in a spine; modified to form a spine.

Spiral – of arrangement, when plant parts are arranged in a succession of curves like the thread of a screw or coiled in a cylindrical or conical manner.

Spiral leaf arrangement (or *alternate leaf arrangement*) – one leaf per node.

Splash-cup (*sporangia*) – a cup-like structure in fungi such as *Nidulariaceae* and cryptogams such as some mosses. The cups function in spore dispersal, in which the energy of raindrops falling into the cup causes the water to splash outward, carrying the spores.

Spongy mesophyll – round, roughly packed cells.

Sporangium (*sporangia*) – a structure in which spores are formed, and the mature spores are released.

Sporangiophore – an organ-bearing sporangia (e.g., the cones of *Equisetum*).

Spore – a haploid (1N) propagule produced by meiosis in diploid (2N) cells of a sporophyte that can germinate to produce a multicellular gametophyte.

Sporocarp – a fruiting body containing spores.

Sporogony – moss sporophyte.

Sporophyll – in pteridophytes, a modified leaf that bears sporangium or sporangia.

Sporophyte – the diploid (2N) multicellular phase in alternating plant and algae generations that produce spores. Compare *gametophyte*.

Copyright © Sterling Education

Sport – a naturally occurring variant of a species, not usually present in a population or plant group; a spontaneously mutated plant differs from its parent plant.

Spreading – extending horizontally (e.g., branches); at right angles to an axis (e.g., leaves or hairs).

Sprout (or *germinate*) – produces buds or branches, a process of growth or development.

Spur – 1. a short shoot. 2. a conical or tubular outgrowth from the base of a perianth segment, often containing nectar.

Squamule (pl. *squamules*, and *squamulae*; adj *squamulose*) – small scales; In lichens, overlapping plate-like forms, sometimes overlapping so much as to become leaf-like, but which lacks a lower cortex, unlike the leafy forms of foliose lichens.

Squamulose – covered with tiny scales (squamules). In lichens, being composed of squamules.

Squarrose – having tips of leaves or stems, etc., radiating or projecting outward (e.g., in the moss *Rhytidiadelphus squarrosus*).

Stalk – the supporting structure of an organ, usually narrower in diameter than the organ itself.

Stamen (adj. *staminate*) – the male organ of a flower, consisting of (usually) a stalk called the filament and a pollen-bearing head called the anther.

Staminate flower (or *male flower*) – a flower with stamens but no pistil.

Staminode – a sterile stamen, often rudimentary, sometimes petal-like.

Staminophore – a structure around the apex of eucalypt (*Myrtaceae* hypanthia) that supports the stamens.

Standard – the large posterior petal of pea-flowers.

Standard specimen – a representative specimen of a cultivar or other taxon that demonstrates how the name of that taxon should be used.

Stele – configuration of vascular tissues in stem or root. *The primary vascular system (including phloem, xylem, and ground tissue) of plant stems and roots.*

Stellate – star-shaped.

Stem – an axial organ of a shoot. The plant axis, either aerial or subterranean, bears nodes, leaves, branches, and flowers.

Stem apical meristem (SAM) – an undifferentiated meristematic tissue in the buds and growing tips of plants. Its primary function is to trigger the growth of new cells in young seedlings at the tips of shoots and forming buds.

Stem-clasping – see *amplexicaul.*

Stenospermocarpy – the development or production of a seedless fruit with minute seeds because of the abortion of seed development. Compare *parthenocarpy*.

Any duplication (copies, uploads, PDFs) is illegal.

Sterile – infertile, as with a stamen that does not bear pollen or a flower that does not bear seed.

Stigma – the top of a female flower part; collects pollen; the receptive surface for pollen. The pollen-receptive surface of a carpel or group of fused carpels, usually sticky, is a point or small head at the summit of the style.

Stipe – a small stalk or stalk-like structure. The stalk of a frond of a fern; the stalk supporting the pileus of a mushroom; the stalk of a seaweed such as a kelp; the stalk-like support of a gynaecium or carpel.

Stipella (or *stipel;* pl. *stipellae*) – one of two small secondary stipules at the base of leaflets in some species.

Stipitate – stalked; borne on a stipe; of an ovary, borne on a gynophore.

Stipulate – bearing *stipules.*

Stipules – small attachments to the leaf, typically located near the base of the petiole. Small appendages at the bases of leaves in many *dicotyledons.*

Stock – see rootstock.

Stolon (or *runner*) – aboveground horizontal shoot. A slender, prostrate or trailing stem, producing roots and sometimes erecting shoots at its nodes. See *rhizome.*

Stoloniferous – having *stolons.*

Stomata (or *stomates, stoma*) – tiny pores opened and closed by guard cells in the underside epidermis that allow water, oxygen, and carbon dioxide to pass into and out of a plant.

Stone cell – a sclereid cell, such as the cells that form the tissue of nut shells and the stones of drupes.

Striate – striped with parallel, longitudinal lines or ridges.

Strigose – covered with appressed, straight, rigid, bristle-like hairs, the appressed equivalent of hispid.

Strobilus (pl. *strobili*) – a cone-like structure consisting of *sporophylls* (e.g., conifers and club mosses) or *sporangiophores* (e.g., in Equisetopsida) borne close together on an axis.

Style – an elongated part of a carpel or a group of fused carpels between the ovary and the stigma. The part of the female flower that connects the stigma to the ovary. *Pollen travels down the style to reach the ovary, where fertilization occurs.*

Stylodium – an elongated stigma resembling a false style (e.g., common in *Poaceae* and *Asteraceae*).

Stylopodium – a swelling on top of the ovary at the base of the styles commonly found in flowers of the *Apiaceae.*

Stylulus – the elongated apex of a free carpel that functions like the style of a syncarpous ovary, allowing pollen tubes from its stigma to enter the locule of only that carpel.

Subcoriaceous – slightly leathery or coriaceous.

 Copyright © Sterling Education

Subgenus – a category of supplementary taxa intermediate between genus and section. The name of a subgenus is a singular noun, always has a capital initial letter, and is combined with the generic name (e.g., *Primula* subgenus *Primula*).

Subglobose – inflated but less than spherical. See globose.

Subshrub (or *undershrub*) – a small shrub that may have partially herbaceous stems but generally a woody plant less than 1 meter (3.3 ft) high.

Subspecies – a taxonomic category within a species, usually used for geographically isolated or morphologically distinct populations of the same species. Its taxonomic rank occurs between species and variety.

Subtend – to stand beneath or close to, as in a bract at the base of a flower.

Subquadrangular – not quite square. Compare *quadrangular*.

Subulate – narrow and tapering gradually to a fine point.

Succulent – plants that accumulate water. 1. juicy or fleshy. 2. a plant with a fleshy habit.

Sucker – a shoot of more or less subterranean origin; an erect shoot originating from a bud on a root or a rhizome, sometimes at some distance from the stem of the plant.

Suffrutex (pl. *suffrutices*) – a *subshrub* or *undershrub*.

Sulcate – furrowed; grooved. May be single (*monosulcate*), two (*bisulcate*), or many (*polysulcate*).

Superficial – on the surface.

Superior ovary – an ovary borne above the level of attachment of the other floral parts or above the base of a hypanthium. Compare *inferior ovary* and *half-inferior ovary*.

Surface/volume law – when body size grows, body surface grows slower than body volume (and weight).

Suspended – of an ovule, when attached slightly below the summit of the ovary. Compare *pendulous*.

Suture – a junction or seam of union. See *fissure* and *commissure*.

Sward – extensive, more or less even cover of a surface (e.g., lawn grass). Compare *tussock*.

Sympatric – having more or less similar or overlapping ranges of distribution.

Sympodial – a mode of growth in which the main axis is repeatedly terminated and replaced with a lateral branch (e.g., family *Combretaceae*, including the genera Terminalia and Combretum). Compare *monopodial*.

Syconium – a hollow infructescence containing multiple fruits (e.g., fig).

Syn– (or *sym*–) (*prefix*) – meaning *with or together*.

Any duplication (copies, uploads, PDFs) is illegal.

Symmetrical – capable of being divided into at least two equal, mirror-image halves (e.g., zygomorphic) or having rotational symmetry (e.g., regular or actinomorphic). Compare *irregular* and *asymmetrical.*

Sympetalous – having united (connate or fused) petals, not free.

Sympodial branching – when the terminal bud degrades every year.

Synangia – adnate sporangia. A fused aggregate of sporangia (e.g., in the trilocular sporangia of the whisk fern *Psilotum*).

Synanthous – growth in which new leaves and flowers appear and die back simultaneously. See *hysteranthous* and *proteranthous*.

Synaptospermy – the dispersal of diaspores as units, where each bears more than one seed (e.g., where each diaspore comprises an entire inflorescence). *True* synaptospermy is when the diaspore remains entire until germination, as commonly happens in species of *Grielum*.

Syncarpous (*of a gynoecium*) – composed of united carpels.

Synonym – an outdated or '*alternative*' name for the same taxon.

Synoecious – a synonym of bisexual.

Syntepalous – having fused *tepals*. See *sympetalous* (i.e., having fused petals).

T

Taproot – the primary descending root of a plant with a single dominant root axis. The large central root extends straight down into the soil.

Tap root system – primary root well developed.

Tartarus – has a coarse, thick, rough, and crumbling surface.

Taxon (pl. *taxa*) – a group or category in a biological classification system.

Taxonomy – the study of the principles and practice of classification.

Tegmen – the inner layer of the testa (i.e., seed coat). It develops from the inner integument of the ovule.

Tendril – any slender organ modified from a stem, leaf, leaflet, or stipule and used by climbing plants to cling to an object.

Tepal – a sepal or petal perianth segment, usually for perianth segments appearing indistinguishable.

Terete (or *semiterete*) – circular in cross-section; cylindrical without grooves or ridges.

Copyright © Sterling Education

Terminal – situated at the tip or apex.

Terminal characters – in leaf description, characters that apply only to the leaf terminals (leaflets).

Ternate – in groups of three; of leaves, arranged in whorls of three; of a single leaf, having the leaflets arranged in groups of three.

Terrestrial – of or on the ground; of habitat, on land as opposed to in water (i.e., *aquatic*), on rocks (i.e., *lithophytic*), or other plants (i.e., *epiphytic*).

Testa – the seed coat.

Tetrad – a group of four; usually used to refer to four pollen grains that remain fused through maturity (e.g., in the *Epacridaceae*).

Tetragonal – square; having four corners; four-angled (e.g., stem cross-sections of herbaceous *Lamiaceae*).

Tetramerous – in four parts, particularly concerning flowers; four parts in each whorl. See *trimerous* and *pentamerous*.

Tetraploid – having four complete sets of chromosomes in each sporophyte cell.

Tetraspore – the asexual spore of red algae. It is so named because each sporangium produces just four spores. See *Rhodophyceae*.

Thalamus (*obsolete*) – 1. a synonym for receptacle. 2. the inflorescence disk of members of the *Asteraceae*. 3. a calyx, as used by Carl Linnaeus.

Thallus – flat, non-differentiated body. The "vegetative" part (part other than sexual fruitbodies) of a lichen has both the fungus (mycobiont) and photobiont.

Theca – one of the usually two synangia in which pollen is produced in flowering plants. It consists of two fused sporangia known as pollen sacs. The wall between the pollen sacs disintegrates before dehiscence, which is usually by a common slit.

Thorn – a sharp, stiff point, usually a modified stem that cannot be detached without tearing the subtending tissue, a spine. Compare *prickle*.

Throat – the opening of a corolla or perianth.

Thylakoids – disk-shaped structures that contain photosynthetic pigments.

Thyrse – a branched inflorescence in which the main axis is indeterminate (i.e., *racemose*) and the lateral branches determinate (i.e., *cymose*).

Timber – the wood of trees prepared for use as building material.

Tissue – a union of cells with a common origin, function, and similar morphology.

Tomentum – a dense covering of short, matted hairs. *It is often used as a general term for bearing an indumentum, but this is not recommended.*

Any duplication (copies, uploads, PDFs) is illegal.

Toothed – having a regularly incised margin.

Torus (or *receptacle*) – the axis of a flower; floral axis; torus; for example, in *Asteraceae*, the floral base or receptacle is the expanded tip of the peduncle on which the flowers are inserted.

Tracheary elements – water-transporting dead cells.

Tracheids – tracheary elements without perforations (openings).

Transmitting tissue – see *pollen transmitting tissue*.

Transpiration – the emission of water vapor from the leaves of plants. *The process of losing water (in the form of vapor) through stomata.*

Transverse – longitudinal section.

Trapeziform – 1. like a trapezium (i.e., a four-sided figure with two parallel sides of unequal length). 2. like a trapezoid (i.e., a four-sided figure or quadrilateral, with neither pair of sides equal), it is sometimes misused as a synonym for *rhombic*.

Tree – a woody plant, usually with a single distinct trunk, generally more than 2–3 meters (6.6–9.8 ft).

Triad – a group of three.

Triangular – planar and with three sides.

Tribe – a taxonomic grouping that ranks between genus and family.

Trichome – in non-filamentous plants, hair-like outgrowth from the epidermis (e.g., a hair or bristle) is sometimes restricted to unbranched epidermal outgrowths.

Trifid – split into three parts. See *bifid*.

Trifoliate – a compound leaf of three leaflets (e.g., clover leaf).

Trifoliolate – see *trifoliate*.

Trigonous – triangular in cross-section and obtusely angled. Compare *triquetrous*.

Trimerous – in three parts, particularly concerning flowers, having three parts in each whorl. See *tetramerous* and *pentamerous*.

Trinerved – having three nerves or veins.

Triplinerved (*of leaves*) – has three main nerves, with the lateral nerves arising from the midnerve above the base of the leaf.

Triporate (*of pollen*) – having three pores.

Triquetrous – triangular in cross-section but acutely angled (with 3 distinct longitudinal ridges). Compare *trigonous*.

Trivalve – divided into three valves. See *bivalve*.

Copyright © Sterling Education

Trivial name – the second word in the two-part scientific name. Compare *specific epithets.*

Trophophyll – a vegetative, nutrient-producing leaf or microphyll whose primary function is photosynthesis. It is not specialized or modified for some other function. Compare *sporophyll.*

Trullate – ovate but angled, as with a bricklayer's trowel; inversely kite-shaped. Compare *rhombic.*

Truncate – cut off squarely, having an abruptly transverse end.

Trunk – the main woody stem of a tree, from which branches grow. *The upright, large, and typically woody main stem of a tree.*

Truss – a compact cluster of flowers or fruits arising from one center, evident in many rhododendrons.

Tuber – enlarged portion of the rhizome. Any of many specialized vegetative underground storage organs. They accumulate food and water or in protection from death by fire, drought, or other hard times. Tubers are well differentiated from other plant organs; for example, informally, a carrot is not regarded as a tuber but simply a swollen root. They differ from the tuber of a sweet potato, which has no particular root-like function. Similarly, corms are not regarded as tubers, even though they are underground storage stems. Tubers store food for the plant and have essential roles in vegetative reproduction. Two types are: 1. *stem tubers* formed by the swelling of an underground stem growing from a root or structures such as underground stolons. Stem tubers produce propagative buds at their stem nodes, forming a seasonal perennating organ (e.g., a potato) and 2. *root tuber* (or tuberoid) that do not form nodes like any standard root.

Tubercle – a small wart-like outgrowth or protuberance of tissue.

Tuberculate – covered in tubercles. See *warty.*

Tuberoid – underground storage organ formed by the swelling of a root; it occurs in many orchids.

Tuberous – resembling a tuber or producing tubers.

Tubular – having the form of a tube or cylinder.

Tufted – densely fasciculate at the tip.

Tunic – the outer covering of some bulbs and corms.

Tunicate (*of bulbs*) – consisting of concentric coats.

Turbinate – shaped like a spinning top or beetroot.

Turgid – swollen with liquid; bloated; firm. Compare *flaccid.*

Turgor (pressure) – cellular water pressure that keeps cells firm.

Tussock – a dense tuft of vegetation, usually well separated from neighboring tussocks, for example, in some grasses. Compare *sward.*

Two-ranked – having leaves arranged in two rows in the same plane, on opposite sides of the branch. See *distichous.*

Any duplication (copies, uploads, PDFs) is illegal.

Tyloses – stoppers for tracheary elements made by parenchyma cells and vessel element "*stoppers*."

Type – an item (usually a herbarium specimen) to which the name of a taxon is permanently attached (i.e., a designated representative of a plant name). It is important in determining the priority of names available for a particular taxon.

Type genus – in nomenclature, a single genus on which a taxonomic family is based.

U

Umbel – a racemose inflorescence in which all the individual flower stalks arise in a cluster at the top of the peduncle and are of about equal length; in a simple umbel, each stalk is unbranched and bears only one flower. A cymose umbel looks like an ordinary umbel, but its flowers open centrifugally.

Umbo – a rounded elevation, such as in the middle of the top of an umbrella or mushroom; a central boss or protuberance, such as on the scale of a cone.

Umbonate – having an umbo, with a conical or blunt projection arising from a flatter surface, as on the top of a mushroom or in the scale of a pinecone.

Unciform – hook-shaped.

Uncinate – having a hook at the apex.

Undershrub – a low shrub, often with flowering branches that die off in winter. Compare *subshrub*.

Understory – plant life growing beneath the forest canopy.

Undulate – wavy and not flat. Compare *sinuate*.

Uniflor – having a single flower (uniflory). Compare *pauciflor* (few) and *pluriflor* (many).

Unilocular – having one loculus or chamber (e.g., the ovary in the families *Proteaceae* and *Fabaceae*).

Uniserial (or *uniseriate*) – arranged in a single row or series. Unbranched. Uniseriate.

Unisexual – of one sex; bearing only male or only female reproductive organs, dioecious, dioicous. See Sexual reproduction in plants.

Unitegmic (*of an ovule*) – covered by a single integument. See bitegmic (i.e., having two integuments).

Urceolate – shaped like an urn or pitcher, with a swollen middle and narrowing top. Examples include pitchers of many pitcher plant genera, such as *Sarracenia* and *Nepenthes* species.

Utricle – 1. a small bladder: a membranous bladder-like sac from the ovary wall, thin pericarp, becomes inflated at maturity, enclosing an ovary or fruit. 2. (*in sedges*) fruit in which the fruit is loosely enclosed from a modified tubular bract.

 Copyright © Sterling Education

V

Vallecular canal – a resin canal coinciding with a longitudinal groove in the seeds of *Asteraceae*. A longitudinal cavity in the cortex of the stems of *Equisetum* coincides with a groove in the stem surface.

Valvate (*of sepals and petals in bud*) – meeting edge-to-edge but not overlapping.

Valve – a portion of an organ that fragments or splits open (e.g., the teeth-like portions of a pericarp) in a split (dehisced) capsule or pod when ripe.

Variant – a plant or group showing some difference from the characteristics associated with a particular taxon.

Variegated – irregularly marked with blotches or patches of another color.

Vascular – referring to the conducting tissues (xylem and phloem) of vascular plants.

Vascular bundle – "chords" made of *xylem* (i.e., inner) and *phloem* (i.e., outer) layers. Tissue in the primary stems of vascular plants consists of specialized conducting cells for transporting water (*xylem*) and assimilating (*phloem*).

Vascular bundles –

Vascular cylinder – hollow cylinder made of xylem (inner layers) and phloem (outer layers).

Vascular plants – Pteridophyta and Spermatophyta.

Vascular tissue –water, nutrient, and photosynthate-conducting tissue (xylem and phloem).

Vasculum – a container used by botanists for collecting field specimens.

Vein (or *nerve*) – a strand of vascular tissue (e.g., in the leaves of vascular plants).

Veinlet – a small vein; the ultimate (visible) division of a vein.

Velamen – absorption tissue made of dead cells. A spongy tissue covering the aerial roots of orchids and some other epiphytes.

Velutinous – see *velvety*.

Velvety – densely covered with fine, short, soft, erect hairs.

Venation – the arrangement of veins in a leaf.

Ventral [Latin, meaning *belly*] – the opposite of dorsal. Partly because the term initially referred to animals rather than plants, usage in botany is arbitrary according to context and source. In general, ventral refers to the belly or lower part, but in botanical usage, such concepts are not always clearly defined and may be contradictory. For example, facing toward the axis (adaxial) refers to an erect plant's lateral organ. Facing toward the substrate in any part of an erect plant, for example, the lower surface of a horizontal leaf (abaxial), the substrate in a prostrate or climbing plant.

Any duplication (copies, uploads, PDFs) is illegal.

Vernation – the arrangement of unexpanded leaves in a bud; the order in which leaves unfold from a bud.

Vernonioid – in the Compositae, style with sweeping hairs borne on abaxial surfaces of style branches.

Verruciform – wart-like in form.

Verrucose – having warts.

Verruculose – minutely verrucose; minutely warty.

Versatile (*of anthers*) – swinging freely about the point of attachment to the filament.

Verticillate – arranged in one or more whorls (i.e., several similar parts arranged at the same point of the axis, e.g., leaf arrangement). Compare *pseudoverticillate* (appearing whorled or verticillate but not so).

Verticillaster – a pseudo verticillate inflorescence typical of the *Lamiaceae*, in which pseudo-whorls are formed from pairs of opposite cymes.

Vessel – a capillary tube formed from a series of open-ended cells in the water-conducting tissue of a plant.

Vessel members – tracheary elements with perforations (openings).

Vestigial – reduced in form and function from the usual or ancestral condition.

Villous – abounding in or covered with long, soft, straight hairs; shaggy with soft hairs.

Vine – 1. scandent plants climbing using trailing or twining stems or runners. 2. such stem or runner. 3. a member of the genus *Vitis*.

Virgate (or *virgulate*) – wand-shaped, twiggy, especially referring to erect, straight stems. In mycology, referring to a pileus with radiating ribs or lines.

Viridiplantae – a clade of autotrophic organisms that includes the green algae, Charophyta, and land plants, all of which have cellulose in their cell walls, chloroplasts derived from primary endosymbiosis with Cyanobacteria that contain chlorophylls a and b, and lack phycobilins.

Viscid – sticky; coated with a thick, syrupy secretion.

Visible spectrum – the resulting array of colors, ranging from red to violet. Mnemonic: ROY G BIV. Each light color has different wavelengths and a different amount of energy.

Viviparous – 1. refers to seeds or fruits germinating before being shed from the parent plant. 2. the development of plantlets on non-floral organs (e.g., leaves).

 Copyright © Sterling Education

W

Warty – a surface covered with tiny round protuberances, especially in fruit, leaves, twigs, and bark. See tuberculate.

Water shoot – an erect, strong-growing, or epicormic shoot developing from near the base of a shrub or tree but distinct from a sucker.

Weed – 1. any plant growing where it is not wanted; commonly associated with disrupted habitats. See *ruderal.* 2. an unwanted plant that grows among crops. 3. a naturalized, exotic, or ecologically *out-of-balance* indigenous species outside of the agricultural or garden context, adversely affecting the survival or regeneration of indigenous species in natural or partly natural vegetation communities because of the invasion.

Wild – originating from a known wild or purely natural habitat (wilderness).

Whorl – flower parts attached to one node. A ring of organs borne at the same level on an axis (e.g., leaves, bracts, floral parts).

Wing – 1. a membranous expansion of a fruit or seed that aids in dispersal, for instance, on pine seeds. 2. a thin flange of tissue extending beyond the typical outline of a structure (e.g., on the column of some orchids, on stems, on petioles). 3. one of the two lateral petals of a flower is located between the adaxial standard (banner) petal and the two abaxial keel petals.

Wood – secondary xylem, stem: everything deeper than vascular cambium.

Woolly – very densely covered with long, more or less matted or intertwined hairs, resembling a sheep's wool.

Work – the ability to change or move matter against other forces.

X

Xeromorph – a plant with structural features (e.g., hard or succulent leaves) or functional adaptations that prevent water loss by evaporation; usually associated with arid habitats but not necessarily drought-tolerant. Compare *xerophyte*.

Xerophyte – a plant adapted to scarce water. A plant living in a dry habitat typically shows xeromorphic or succulent adaptation and can tolerate long periods of drought. Compare *xeromorph.*

Xylem (or *sapwood*) – a specialized water-conducting tissue in vascular plants. Vascular tissue transporting water and nutrient-conducting tissue that carries water and nutrients from the roots to the leaves. Plant tissue that conducts water and dissolved nutrients. *As new layers develop, the inner layers die and become heartwood.*

Any duplication (copies, uploads, PDFs) is illegal.

Z

Zonate – having light and dark circular bands or rings, typically on leaves or flowers.

Zygomorphic – bilaterally symmetrical; symmetrical about one vertical plane only; applies to flowers where the perianth segments within each whorl vary in size and shape.
Contrast *actinomorphic* and *irregular*.

Zygote – a fertilized cell, the product of the fusion of two gametes.

Copyright © Sterling Education

Frank J. Addivinola, Ph.D.

The lead author and chief editor of this study guide is Dr. Frank Addivinola. With his outstanding education, laboratory research, and decades of university science teaching, Dr. Addivinola lent his expertise to develop this book.

Dr. Frank Addivinola conducted original research in developmental biology as a doctoral candidate and pre-IRTA fellow in Molecular and Cell Biology at the National Institutes of Health (NIH). His dissertation advisor was Nobel laureate Marshall W. Nirenberg, Chief of the Biochemical Genetics Laboratory at the National Heart, Lung, and Blood Institute (NHLBI). Before NIH, Dr. Addivinola researched prostate cancer in the Cell Growth and Regulation Laboratory of Dr. Arthur Pardee at the Dana Farber Cancer Institute of Harvard Medical School.

Dr. Addivinola holds an undergraduate degree in biology from Williams College. He completed his Masters at Harvard University, Masters in Biotechnology at Johns Hopkins University, and five other graduate degrees at the University of Maryland University College, Suffolk University, and Northeastern University.

During his extensive career, Dr. Addivinola held faculty positions at colleges and universities, including Harvard University, Johns Hopkins University, University of Maryland and Northeastern University, and taught numerous undergraduate and graduate-level courses in biology, biochemistry, organic chemistry, inorganic chemistry, anatomy and physiology, medical terminology, nutrition, and medical ethics. He received several awards for his research and presentations.

Essential Biology Self-Teaching Guides

Eukaryotic Cell & Cellular Metabolism

Molecular Biology & Genetics

Nervous & Endocrine Systems

Circulatory, Respiratory & Immune Systems

Digestive & Excretory Systems

Muscle, Skeletal & Integumentary Systems

Reproduction & Development

Microbiology

Plants & Photosynthesis

Evolution, Classification & Diversity

Ecology & Population Biology

Visit our Amazon store

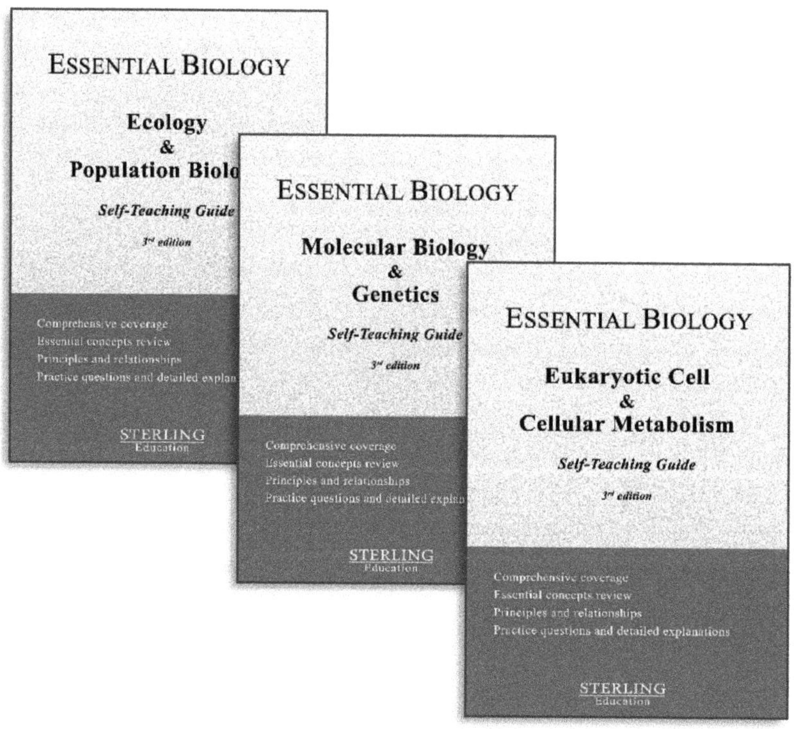

Essential Physics Self-Teaching Guides

Kinematics and Dynamics

Equilibrium and Momentum

Force, Motion and Gravitation

Work and Energy

Fluids and Solids

Waves and Periodic Motion

Light and Optics

Sound

Electrostatics and Electromagnetism

Electric Circuits

Thermal Physics

Atomic and Nuclear Physics

Visit our Amazon store

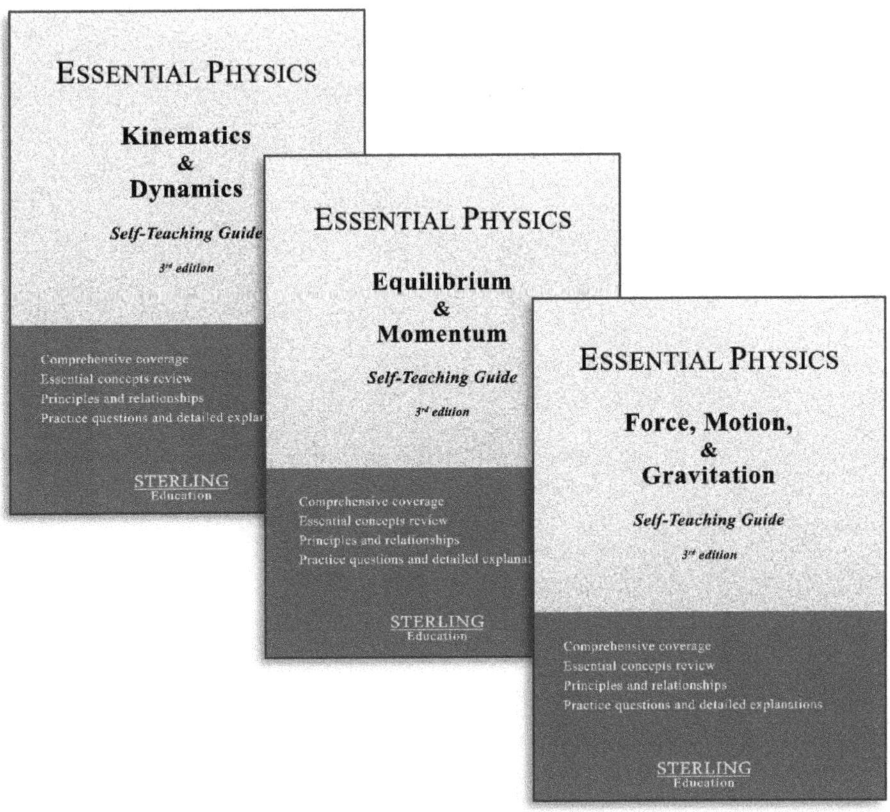

Essential Chemistry Self-Teaching Guides

Electronic Structure & Periodic Table

Chemical Bonding

States of Matter & Phase Equilibria

Stoichiometry

Solution Chemistry

Chemical Kinetics & Equilibrium

Acids & Bases

Chemical Thermodynamics

Electrochemistry

Visit our Amazon store

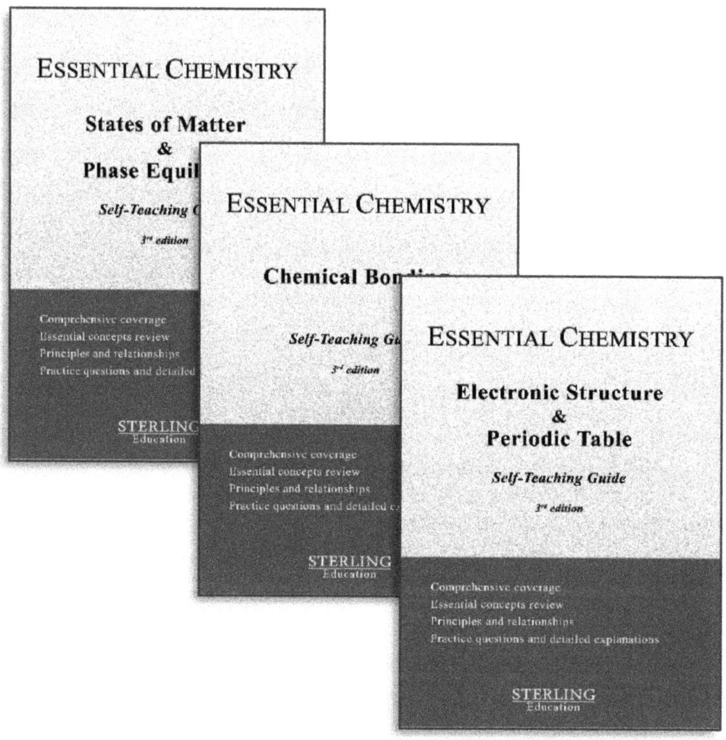

Everything You Always Wanted to Know About…

Chemistry

Physics

Cell & Molecular Biology

Organismal Biology

Human Anatomy & Physiology

Environmental Science

American History

American Law

American Government & Politics

Comparative Government & Politics

World History

European History

Psychology

Sociology

Human Geography

Visit our Amazon store

For online practice resources visit
https://www.sterling-prep.com

If you benefited from this book, please leave a review on Amazon so others can learn from your input. Reviews help us understand our customers' needs and experiences while keeping our commitment to quality.

www.ingramcontent.com/pod-product-compliance
Ingram Content Group UK Ltd.
Pitfield, Milton Keynes, MK11 3LW, UK
UKHW051452290926
15471UKWH00008B/1151

9 798885 572798